T0136077

About Island Press

Since 1984, the nonprofit organization Island Press has been stimulating, shaping, and communicating ideas that are essential for solving environmental problems worldwide. With more than 800 titles in print and some 40 new releases each year, we are the nation's leading publisher on environmental issues. We identify innovative thinkers and emerging trends in the environmental field. We work with world-renowned experts and authors to develop cross-disciplinary solutions to environmental challenges.

Island Press designs and executes educational campaigns in conjunction with our authors to communicate their critical messages in print, in person, and online using the latest technologies, innovative programs, and the media. Our goal is to reach targeted audiences—scientists, policymakers, environmental advocates, urban planners, the media, and concerned citizens—with information that can be used to create the framework for long-term ecological health and human well-being.

Island Press gratefully acknowledges major support of our work by The Agua Fund, The Andrew W. Mellon Foundation, Betsy & Jesse Fink Foundation, The Bobolink Foundation, The Curtis and Edith Munson Foundation, Forrest C. and Frances H. Lattner Foundation, G.O. Forward Fund of the Saint Paul Foundation, Gordon and Betty Moore Foundation, The Kresge Foundation, The Margaret A. Cargill Foundation, New Mexico Water Initiative, a project of Hanuman Foundation, The Overbrook Foundation, The S.D. Bechtel, Jr. Foundation, The Summit Charitable Foundation, Inc., V. Kann Rasmussen Foundation, The Wallace Alexander Gerbode Foundation, and other generous supporters.

The opinions expressed in this book are those of the author(s) and do not necessarily reflect the views of our supporters.

FORESTS IN OUR CHANGING WORLD

Forests in Our Changing World

NEW PRINCIPLES FOR CONSERVATION

AND MANAGEMENT

Joe Landsberg and Richard Waring

ISLANDPRESS

Washington | Covelo | London

Library of Congress Cataloging-in-Publication Data

Landsberg, J. J.
Forests in our changing world: new principles for conservation and management/
Joe Landsberg and Richard Waring.
pages   cm.
Includes bibliographical references.
ISBN 978-1-61091-495-6 (cloth : alk. paper)—ISBN 1-61091-495-3 (cloth : alk. paper)—
ISBN 978-1-61091-496-3 (pbk. : alk. paper)—ISBN 1-61091-496-1 (pbk. : alk. paper)
1. Forest ecology. 2. Forests and forestry. 3. Climatic changes. 4. Trees—ecophysiology.
I. Waring, Richard H. II. Title.
SD390.7.C55L37   2014
577.3–dc23
2013041011

Printed on recycled, acid-free paper ✿

Manufactured in the United States of America

10 9 8 7 6 5 4 3 2 1

Keywords: Island Press, climate, weather, climate change, tree physiology,
forest health, forest types, forest products, forest productivity, forest management,
forest disturbance, forest clearance, forest values, ecosystem services, plantations,
economics, biodiversity, hydrology, computer simulation models,
resilience, wood production

CONTENTS

We have had long careers as forest ecologists—researchers con-cerned with finding out how trees and forests grow. About the time that we started out on those careers the world's human population was around three billion. It's now over seven billion and climbing steadily. Populations have stabilized in most of the developed world but not in sub-Saharan Africa, India, and other parts of Asia. Over-all, the inexorable rise continues toward a predicted maximum of more than nine billion people.

High human populations bring pressure on the world's natu-ral resources. Biological resources are already being consumed far faster than they can be replaced, and the pressure is growing even faster than the increase in human numbers. Modern communica-tions reach the most out-of-the-way places, so billions of people who may have lived simple lives for generations are now aware of the way the rich world lives, and they want to live like that. They want good houses, electricity, running water on tap, and consumer goods of all sorts. They also want to eat better and more varied diets than those they once might have accepted without question. And people in the rich, developed countries, who already have high liv-ing standards, have grown accustomed to the idea that those living standards should continue to improve.

Many people are aware of the implications of our massive resource use and the impacts it's having on the climate of our earth. Some are aware of the widespread land degradation and the pollution of freshwater, the oceans, and the air. But relatively few are aware of the impacts of the demand for resources on forests, or of the implications of those impacts. Forests are just "out there"—part of the scenery. Most people, if they think about it at all, are aware that the wood used by our societies comes from forests of some kind, but they don't know much (if anything) about the different types of forest, how much of it there is, and how it's managed. That's all done, presumably, by people who know about forests and how to work in and manage them.

Those people are the professional foresters. They're concerned with the wood yields that can be obtained from particular areas of forest, how to get the wood out and how to manage those forests so that they will continue to produce wood. Most professional foresters are good conservationists: they're not interested in destroying the forests. But, at least until recently, most professional foresters were not particularly interested in the physiological processes that determine how trees grow. There wasn't much to indicate how knowledge about those processes contributed to their ability to make decisions about how to manage the forests.

That's changing now. We have argued, for many years, that it's important to understand how interactions between physiological processes and climatic conditions influence the growth of trees, and we have described in detail in other books the contribution that knowledge of ecophysiological processes can make to management decisions. In this book we describe those processes, and explain their significance, in relatively simple terms. On the first occasion that we use any technical term, with which readers may be unfamiliar, we italicize the word and define it. A list of those terms and their definitions is provided in the glossary at the end of the book. Our intention is to provide a framework on which to build the detailed recipes that will be needed in particular cases and circumstances.

The fact that the world's climate is changing makes it even more important that we use knowledge about tree physiology and forest/weather interactions to guide management decisions and forest policy. We also need to use that understanding to help promote appreciation of the devastating impacts of the widespread forest destruction that is being (largely) driven by commercial pressures. It's essential that as many people as possible, from all walks of life, understand the importance of the choices that have to be made in relation to forests, and the natural environment in general.

In aiming to reach a wider audience, improve public knowledge about forests, and help generate support for the policies and practices that are essential to sustain a resource vital to human life on earth, we are also aiming to strengthen the case for saving the world's natural forests. That doesn't mean they can't be used: we can "have our cake and eat it too." High-value timber can be harvested without destroying the systems that produce it, while most of the wood for pulp and timber for building can come from plantations, which are simplified forests. But forests aren't just systems for providing wood or improving the scenery. They can be managed to provide the ecosystem services that we need even more than we need timber. Those services have real, material value, which must be included in making decisions about the way we use—or misuse—forests.

Scientists are supposed to be coldly logical and objective, and of course we do our best, but we also have to say that we subscribe to the sentiments attributed to Chief Seattle: "This we know: the earth does not belong to man, man belongs to the earth. All things are connected like the blood that unites us all. Man did not weave the web of life, he is merely a strand in it. Whatever he does to the web, he does to himself." Forests are part of the earth's web of life. No amount of material consumerism will be able to compensate for their loss, or even their degradation.

On Units and Notation

Metric units are universal in science and are in everyday use in every country except the United States. The system is officially accepted there but has not been adopted as the official system of measurement. Some of our readers may, therefore, be more familiar with the imperial system of measures. There is no point in providing a large table of conversions; they can be looked up on any number of websites, or even in the front of books such as desk diaries. However, for convenience, we note here the conversions from the metric units that we use most frequently, to imperial units:

1 kilogram (kg) = 2.2 pounds (lb.)
I metric ton (t) = 1,000 kg = 2,205 lb.
1 meter (m) = 3.28 ft.
1 cubic meter (m^3) = 35.31 $ft.^3$
1 hectare = 100 x 100 m = 10,000 m^2 = 2.47 acres

We use scientific notation throughout, that is, instead of writing *kilograms per hectare per day*, or *kg/ha/day*, we write *$kg\ ha^{-1}\ day^{-1}$*. The –1 exponent/superscript indicates that the variable in this case is a divisor. This is universal in science because it makes the manipulation of numbers, and analysis of dimensions, a great deal simpler.

For example, to multiply 500 kg ha^{-1} day^{-1}, which we can write as 5 x 10^2 by, say, 300 days (3 x 10^2), we see that the total is 15 x 10^4 (adding the exponents) and, since we have (kg [ha x days]) x days, the days cancel out and the answer is 15 x 10^4 kg ha^{-1}, which is 150,000 kg ha^{-1}, or 150 t ha^{-1}.

Similarly, if we're thinking of water in the landscape, we note that 1 liter of water (1,000 milliliters, ml) spread over 1 square meter (1 m^2), which is 100 x 100 cm^2 = 10^4 cm^2, is (1 x 10^3)/(1 x 10^4) ml/cm^2. That is, (1 x 10^3) x (1 x 10^{-4}) ml cm^{-2}. Adding the exponents gives 1 x 10^{-1}, which is 0.1 ml cm^{-2}. Since 1 ml = 1 gm, and 1 gm = 1 cm^3, 0.1 ml = 0.1 cm^3, which is equivalent to a depth of 1 mm. This is one of the very convenient outcomes of the metric system. Extending it to larger areas: 1 ha = 1 x 10^4 m^2, so if 1 mm of rain falls on 1 ha it is equivalent to

1 liter m^{-2} = 10,000 liters. Very useful in hydrology!

Note that temperature is expressed in degrees Celsius, °C (centigrade), throughout.

Joe Landsberg, *Mount Wilson, New South Wales, Australia*
Dick Waring, *Corvallis, Oregon*

ACKNOWLEDGMENTS

There is, we believe, widespread interest in forests and global warming, among people who are not foresters. So we aimed this book at undergraduates and interested "lay" people (in relation to forests and forest ecology). This imposed on us the need to write in a style different from the dry and rigorous scientific style we are accustomed to. We had to be a bit more relaxed, and we have also had to present our ideas and materials differently.

We needed someone to give us some feedback (the word is in the glossary) on whether we have succeeded. For this we are very grateful to Joe's daughter Sue Moorhen—mother, businesswoman, and engineer—who found the time in her busy life to read the manuscript (well, almost all of it) and offer detailed comments and critique from the point of view of a technically educated, environmentally concerned, non-forester and non-biologist. We didn't always follow her advice (authors never do), but it was a considerable and much-appreciated effort. Joe's wife, Diana, also read the whole manuscript and offered feedback from the point of view of an educated—but decidedly nontechnical—person. A labor of love. She told us if she didn't understand what we were trying to say and provided useful comments and suggestions on grammar (most of which we accepted).

For help and comment on all aspects—technical and relating to how we expressed ourselves—we are grateful to friend, colleague, ecophysicist, and objective critic Dr. Peter Sands, who worked though the drafts, chapter by chapter, and provided his usual helpful and accurate suggestions. It's not Peter's fault if something silly has slipped through. Greg Heath produced all the diagrams, Professor Sune Linder read and commented on several chapters and provided valuable help with the reference list. Dr. Merrill Kaufmann provided suggestions about fire.

Chapter 1

Introduction: Looking Back and Into the Future

Our objective in this book is to describe and discuss forests and their significance in our world. Human societies need the products of forests—not just wood and wood products but all the ecological goods and services that forests provide: biodiversity and its essential benefits, carbon sequestration and storage, stable water supplies, land protection, recreation. But relatively few people are aware of these services and benefits, so we hope to contribute to raising awareness of these values and the importance of forests, and to providing the science-based information needed to guide political action and decisions about them. Toward achieving these objectives we consider how forests grow and why different types occur in different parts of the earth; what constrains their growth, why they are important to us and how they should be managed.

Most people like trees. We plant and nurture them in our parks and gardens—in fact the very idea of a suburb suggests leafy, tree-lined streets—but there are big differences between trees in the suburbs or scattered around farmhouses or in small woodlots, and those in forests. Forests are embedded in our psyche: they have been important to us throughout our history, but most people, nowadays, know too little about their importance to the planet and to our lives and economies.

People of different cultures and backgrounds view forests in different ways: some see them as rather mysterious wilderness, others simply as blocks of land with commercial potential. Forests are basic to the folklore of people who have lived in and with them for generations. Scandinavian and German myths and legends tend to involve dark and sometimes forbidding forests. In some countries, access to the forests to camp, collect berries and mushrooms, hunt, or simply to walk in and enjoy them is a right entrenched in law.

In the modern world virtually all forests are exploited by humans in various ways, mostly, of course, for wood production for industrial purposes or for fuel. Some natural forests are protected and well managed but economic pressures are bringing about the destruction of many others. Tropical forests provide a living to—sadly, remnant—native peoples in Indonesia, Papua New Guinea, and the Amazon, but those forests are being destroyed at frightening rates by illegal logging or to establish oil palm plantations or cleared for crop production. Forests have been and, indeed, remain among the dominant vegetation types across the earth, but one of the most important processes by which humans have transformed the earth is through deforestation. In the following paragraphs we provide an outline of the history of human interaction with forests and a short synopsis of the history of deforestation around the world. In later chapters we consider some of the consequences and implications of destroying forests.

Forests in Human History

It's now generally accepted by anthropologists and archeologists that our species evolved in the forests of central Africa somewhere around 3.5 million years ago—less than 1 percent of the total age of our earth. Why those early ape-like creatures started to walk upright on two limbs instead of just going on the way they were, presumably swinging through the trees, is a matter of speculation. It seems inarguable that our very early ancestors moved from living in the central African forests to the savannah; the species called *Homo sapiens* emerged about half a million years ago and became essentially identical to modern humans about 50,000 years ago.

Much of the early development of human societies was in the area covered by modern Israel, Lebanon, Syria, Iraq, and eastern Iran, none of which was heavily forested. From there, early humans radiated west along the north coast of the Mediterranean Sea and east, within the warm subtropical latitudes. There is also archeological evidence of *Homo erectus* (not yet *H. sapiens*) activity in China more than a million years ago.

It's estimated that the human population of the earth has increased from a few hundred thousand at the time of the retreat of the glaciers, around 10,000 years ago, to about 200 million at the beginning of what we in the Western world call the Christian era. Of that 200 million, a significant proportion was in China. Morris (2010) says there is evidence that rice and millet were cultivated in the Yangtze valley between 8,000 and 9,000 years ago, and that 6,000 years ago the Yangtze and Yellow River valleys were mostly subtropical forest. Forests were cleared as the steady growth of human populations led to the expansion of cropland around more and more villages.

As the glaciers and ice sheets of the Ice Age retreated, forests expanded into huge areas where tree growth had previously been impossible. Human populations also increased as the ice retreated and people moved north into regions of Europe that had been uninhabitable. Human disturbance of forests in Europe seems to have become significant around 6,000 years ago—approximately the same time as in China—and increased from then on. There are written accounts from the ancient Mediterranean civilizations, notably Greece and Rome, of extensive forest clearing for wood for fuel, building, and shipbuilding, leading to denudation of the countryside. The damage was exacerbated by widespread overgrazing and browsing, particularly by goats. This led to soil loss by erosion and reduced agricultural productivity, which may well have been a significant contributor to the decline of those civilizations.

The Roman Empire at its peak, near the beginning of the Christian era, included about sixty million people; Rome itself reached about one million, a massive city population for the time. Wood was the most important building material and, throughout the empire, trees were cut for housing and the great shipbuilding program of the Romans, as well as to provide fuel for domestic heating, iron-working, and ceramics manufacture. The expansion of agriculture also resulted in increasing land clearance, so the Roman period saw considerable changes in forest cover, particularly in southern Europe. Clearance in Europe was checked by declining populations associated with the disintegration of the Roman Empire, the invasions by Huns and Goths from the east, constant war, the spread of various lethal diseases, and famine. Populations recovered and increased between the fourth and seventh centuries, with accelerated land clearance; by the end of the fourteenth century "farmers had plowed up vast tracts of what had once been forest, felling perhaps half the trees in western Europe" (Morris 2010, 367). The plagues known as the "black death" caused human populations to crash in the fourteenth century, and large areas of forests recovered by natural regeneration during that period.

Morris (380) also describes how, in the Chinese city of Kaifeng on the Yellow River, iron output increased sixfold between AD 800 and 1100: "Foundries burned day and night, sucking in trees to smelt ores into iron; so many trees, in fact, that ironmasters bought up and clearcut entire mountains. . . . There was simply not enough wood in northern China to feed and warm its million (human) bodies and keep foundries turning out thousands of tons of iron." This demand triggered the increased utilization of a new fuel source—coal. Morris does not concern himself with environmental impacts, although he does comment that "Kaifeng was apparently entering an ecological bottleneck," by which he means, presumably, that it was reaching a tipping point, where the surrounding ecosystems would collapse. In the modern era, forests in China were devastated during Mao Zedong's Great Leap Forward when, to reduce China's need to import steel and machinery, people were encouraged to set up backyard steel furnaces to turn scrap metal (including their own pots, pans, and farm implements) into steel. This resulted in very little, if any, usable steel, while entire forests were cut down to fuel the smelters, leaving the land vulnerable to erosion and contributing to massive environmental damage.

Discussing the impact of humans on the land in the immediate preindustrial period, Morris (2010) tells us, "One scholar complained in the 1660s that four-fifths of Japanese mountains had been deforested." The same thing happened in Britain. The forests of Britain had been cleared for farmland well before the arrival of the Romans[1] and later were heavily exploited for timber; only 10 percent of England and Scotland were still wooded around 1550, and by the 1750s most of those trees were gone too. The demand for timber, and consequently much of the deforestation in England between the sixteenth and eighteenth centuries, was caused by shipbuilding: Britannia, as it was happy to tell the world, ruled the waves, and to do that it needed ships. The British navy was the means by which the island kingdom protected itself and projected its power; thousands of wooden ships were built through the seventeenth, eighteenth, and nineteenth centuries. The toll on the forests was enormous. English demand for access to American forests and wood supplies was one of the sources of tension between England and its American colonies in the eighteenth century. Ireland, by contrast, was still 12 percent forest in 1600, but colonists eliminated more than 80 percent of those trees by 1700.

In Europe the need for timber drove exploitation of the forests, but the need to manage them in a systematic way that would ensure their survival and long-term productivity (today we would use the word *sustainability*) was recognized early, particularly in Germany, where formal forest management dates back to the fourteenth century. Germany was not a single country at that time, so management practices and control of the forests varied between states. Some of the mixed temperate forests in Germany are regrowth; they were at one time completely destroyed, as in the northeastern United States (see our comment a little later, about the recovery of those forests).

In his fascinating and salutary book *Collapse*, Jared Diamond (2005) examines the reasons various human societies of the past have collapsed. They are as varied as the societies he considers, but a common thread is environmental destruction and deforestation, leading to degeneration of the ecosystems on which every society relies, and ultimately the collapse of the societies. Deforestation was seldom the only problem, although in the case of Easter Island, where trees provided the rollers to move stone statues, it was the primary cause of soil erosion, loss of soil fertility and productivity, and the loss of material to make canoes for fishing and trading. Eventually deforestation brought about the total disintegration of the Easter Island society. In the highly successful Mayan civilization on the Yucatan peninsula, which lasted from about AD 250 to 900, high population growth outstripped the availability of resources. The process was accelerated by deforestation of the hillsides, leading to soil erosion and the accumulation of infertile sediment in the valley bottoms and loss of agricultural production. In modern times the island of Haiti, in the Caribbean, is a failed state with unstable government and frequent breakdowns of law and order. Heavily overpopulated and desperately poor, it has been almost completely deforested with disastrous effects on water supplies and agricultural production. Things are very different in the Dominican Republic, which occupies the other section (two-thirds) of the island. Forest cover there is good and the ecology seems to be healthy. Perhaps not coincidentally, the Dominican Republic is economically healthy and politically stable.

A point we need to make with respect to all this historical deforestation is that it was rarely deliberate, at least not by native inhabitants. People needed wood and fuel and they needed to clear land for agriculture and living space. The forests, in many cases, seemed huge and a few people with axes would not have felt they were doing significant damage. Each generation was comfortable with its actions and did not consciously set out to destroy the forests. In most cases each generation had increasing appreciation for a declining resource. Even in Australia, where permanent land settlement only started in the nineteenth century, the forests seemed endless and men with axes and cross-cut saws, wielded with immense labor, could not imagine that they would end up destroying a large proportion of them. Things are different now. With our communication systems, satellite coverage of the earth, and methods of disseminating knowledge, we have no excuse for not knowing what's going on. We know how much forest there is, how fast it grows, and how fast we are harvesting or destroying it. We understand the importance of forests and what their loss entails. In this book we discuss what we need to know to use forests sensibly, so we can benefit from them—as humans always have—without destroying them.

But across the world, as we write, the destructive trend continues. Amazonian rainforests are being destroyed to make way for commercial production of soybeans, mainly for export to the United States. Indonesian and Malaysian rainforests are cleared to free land for agro-industrial purposes, such as oil palm cultivation. The rainforests of Papua New Guinea, the Solomon Islands, Vanuatu, and other Pacific islands are being destroyed to satisfy voracious Korean, Japanese, Taiwanese, and Chinese commercial interests. Poor decisions, such as those based on superficial economic analyses, without taking account of the long-term implications, can lead to irreversible loss of options in the future. For example, a recent analysis of the effects of climate change on forest production in California notes that pulp yields (for paper production) will decrease, but the land values for urban development will increase. Therefore, in conventional economic terms, additional urban development (sprawl) is considered justifiable: the gain in land values is deemed to compensate for the loss in forests and forest productivity, although the environmental costs are likely to be high. We discuss the values of forests and the economics of using them in chapters 5 and 6.

The forest cover of Australia has been estimated at 33 percent at the time of European settlement. It is now 19 percent. The country has a remarkably cavalier attitude about its forest resources: clear-felling (clearcutting) and logging of old-growth forests continue to this day, although this activity has now been greatly reduced, largely because of pressure brought by an increasingly concerned public. But much of Australia's agricultural land—land that was covered by forests, or at least woodlands—now has serious salinity problems, because the deep-rooted trees that were once present, transpiring water all the year round and keeping the water tables down, are largely gone.

There are some bright spots in all this. A forest is not necessarily destroyed when all the trees are cut down. It is destroyed when the land is no longer allowed to support trees, but is converted to crops, pastures, or urban development. If once-forested lands are left alone after clear-felling, the forests will, in most cases, regenerate and reestablish themselves, sometimes with astonishing rapidity. The trees may not initially be the same species but under a stable climate will evolve to once more resemble a natural forest. We will discuss the processes involved in more detail later.

The United States provides examples. It was colonized from east to west by Europeans, who in the seventeenth and eighteenth centuries cleared the forest from large areas of New England for agriculture. As the colonists moved west and the more productive and easily farmed lands of Ohio, Indiana, Illinois, and Iowa came into agricultural production, and transport improved, it became uneconomical to farm in New England and New Hampshire, so the cleared lands were abandoned. These areas are now thriving hardwood and mixed forests. Similarly, large areas in the southern United States were cleared of forests to grow cotton. These too were abandoned and forests reestablished themselves across millions of hectares. Large areas of denuded cotton lands have also been planted to conifer plantations, which, with appropriate management to deal with the results of soil degradation by the exploitive cotton culture, are highly productive. They are not always native trees, but plantations are part of the "family" of forests. We will discuss them in some detail later. In Russia, large areas of cooperative farmland have been abandoned since the fall of the Soviet Union in 1989, and abundant regrowth of forests is now evident.

But, despite the few bright spots, and regeneration, overall the world is losing its forests at a depressing rate. This is not just the result of land clearance. All natural ecosystems are under serious stress as a result of the pressures generated by the demands and activities of more than seven billion humans. Fossil fuel burning has raised the carbon dioxide (CO_2) and other greenhouse gas concentrations in the atmosphere to levels that have not been seen for at least the last two to three million years, leading to rising global temperatures, which, in some areas, are causing prolonged and unusual droughts (chapter 4). Water shortage in itself may kill trees, and drought-stressed trees are more likely to burn, so forest fires are becoming more frequent and more damaging. There have always been forest fires, but the intensity of the fires that we are seeing now in California and Texas, Spain and Portugal, and in Australia, is frequently higher than in the past, so the trees are killed and some of the forests do not recover. In terms of the concept of ecosystem resilience now becoming widely adopted (see the next section) these forests reach thresholds where the fires cause collapse of the system so that it cannot revert to its previous state, with serious implications for various systems that depend on, and interact with, the forested lands. Rising global temperatures also bring about changes in the lifecycles of insects, including those that damage trees. The result, in some places, is unprecedented damage, which kills whole forests directly. Dead trees are also, of course, much more likely to burn.

So there are plenty of problems. But there's not much point in wringing our hands about them, and succumbing to despair. If any problem is to be solved, the first step is to recognize it, establish what causes and constitutes the problem and then determine how, at least in principle, it can be solved. In the case of forests, the first steps must be to identify the different types, where they occur, and what threats they face. We also need to identify the values that humans place on forests and the products they expect to get from them. These include wood, water, the preservation of biodiversity, and value as recreational areas in which people can hike and camp and that can be enjoyed for their peace and beauty. In the western United States, northern Europe, and Russia—and no doubt other places— forests are also hunting grounds. We want to conserve forests, to manage them sustainably for their value to us as humans and to all the other life forms that depend on or are affected by them, which means managing them so they not only do not deteriorate but, where possible, they are improved. In very few cases will forests be "locked up"—preserved as untouchable areas to which human access is forbidden.

Resilience

We will mention, at various points in this book, the concept of *resilience*. This is becoming widely recognized among scientists, particularly ecologists, as an extremely important property of complex, adaptive, so-called self-organizing systems. Resilience describes the ability of a system to absorb disturbance and still retain its basic structure and ability to function. A key concept in considering resilience is the idea of a threshold, or *tipping point*. This is the point at or beyond which the system will move into a new stable state, that is, one from which it will not return to its previous stable state. The transition from one stable state to another is invariably a nonlinear process; the change may be fast or slow and may involve different scales within the system.

Resilience includes consideration of the needs of humans and socioeconomic systems as well as that of the ecological or agricultural systems. So when we evaluate the resilience of forests we're not considering them as ecological systems quarantined in some way from human activity and responding only to the pressures imposed by nature: we have to consider the purpose for which land is being used and the ability of the forest to maintain its stability as an ecological system while subject to the pressures and disturbances imposed by humans. If human activity causes a forest to "flip" into a new—probably degraded—state from which it might recover if subsequently left undisturbed, but the pressures that caused the change are not removed, then the new state will become permanent.

Examples that we will come across in this book include the impacts of insect attacks on forests, the impacts of fire, and of logging. If serious insect attacks occur in conjunction with disturbing factors such as drought and fire, it is possible some forest stands will be flipped into another stable state, such as a grass- or shrub-dominated system. Many of Australia's eucalypt forests can tolerate, indeed need, relatively low-intensity fires, but if fire frequency is too low because of fire suppression policies, fuel builds up on the forest floor and, at some point during drought and hot, dry weather, there is likely to be a catastrophic fire that kills mature trees and may cause transition to another stable state, very different from the original forest. The same applies to some forests in the western United States. Similar arguments apply to logging. Many forests can absorb the impact of selective logging, in which only a proportion of the trees are removed. But they may not be able to absorb the impact of clear-felling and burning the logging residues on-site. Tropical rainforests can be relatively easily pushed past their tipping point by *slash and burn agriculture* because tropical soils are inherently infertile: the soils are degraded by this form of agriculture so it is unlikely that the forest will regrow. We will consider these and other examples in more detail later. In the chapter on climate change we consider (at least in passing) whether the resilience of the planet as a whole is being seriously tested.

Systems and subsystems are linked; there are feedbacks and interactions even if we can't anticipate all of them and account for the likely effects. In assessing the likely resilience of an area of forest of a particular type we consider the form of disturbance (e.g., management, including logging and clearing or possible insect attack or disease) to which it is, or might be, subject. It's also important to consider the factors, such as soil type, rainfall regime, and characteristics such as current health and biodiversity, which are likely to affect the system's ability to tolerate disturbance. Interacting components, such as understory characteristics and nutrient cycling and turnover, also need to be taken into account. They may affect the response time of the system—the rate of the transition from one state to another—if the system does flip into a different state. Spatial factors can also play a role. The dynamic behavior of small patches of forest that interact with one another (patch dynamics) may determine the resilience of larger areas. Clear-felling a small patch has effects on the surrounding forest; the state of that forest, in turn, affects the capacity of the cleared area to recover. (For those interested in a more detailed exposition of resilience and its implications in our modern world, two clear and readable books by Walker and Salt, *Resilience Thinking* (2006) and *Resilience Practice* (2012), provide an excellent starting point.)

Outline of the Book

We have, so far, talked about forests in a generic sense, assuming we all know what constitutes a forest. But what exactly are they? To many people, particularly city dwellers, forests are areas of country covered by generally large trees with lots of foliage. To foresters concerned with wood production, the forest is an economic unit, to be managed as such. Monospecific plantations are important examples. To ecologists forests are diverse ecological communities of plants that happen to be dominated by trees with closed canopies, which means that the tree crowns meet and overlap. This doesn't necessarily mean complete closure—there will usually be gaps—although the canopies of some very dense forests may prevent virtually all light from reaching the ground. We don't need to be too prescriptive about this: there is a gradation from forests through woodland, to open woodland, savannah, and grasslands, but in general we are talking about systems that are covered by our definition, that is, plant communities dominated by trees, the crowns of which overlap.

Most people are aware, if they travel or watch nature programs on TV, that there are different kinds of forests. In fact there are many types, ranging from tropical forests containing multiple layers of vegetation to boreal forests with relatively simple stand structure and few species. In tropical rainforests large trees of different species create the overarching foliage canopy, beneath which there is a range of understory plants—small trees, shrubs, herbaceous plants—adapted to varying light environments. Most tall forests have understory vegetation, but there are forests dominated by one or two species that inhibit the development of understory, either because the main canopy intercepts virtually all the light, as is the case in some dense conifer forests, or because the large trees of the main canopy use virtually all the available water and create an environment so harsh that few other species can thrive there.

In chapters 2 through 6 we discuss in more detail many of the points we have touched on in this introduction. The main forest types, their characteristics and the differences between them, are outlined in chapter 2, where we also consider tree plantations. There's a good argument to be made that plantations are simply tree cropping systems, with relatively little in common with natural forests. However, we consider plantations to be at the end of the spectrum of forest types: simplified, monospecific, even-aged communities of trees generally planted and maintained to meet the needs of humans for forest products. Plantations are important: they are present on less than 3 percent of forested land, but they produce 25 percent of the wood currently harvested around the world.

There is far more land in the Northern than the Southern Hemisphere, so there are much greater areas of forest. Consequently, much of our discussion concerns forests in the northern half of the world. In general, in the cold and difficult high-latitude regions—the boreal zones—of the Northern Hemisphere the dominant trees are tough, slow-growing conifers, together with a few broad-leaved deciduous species. In milder regions farther south conifers are, again, the dominant species, but these are generally much larger trees; the huge spruce, firs, and redwoods in the Pacific Northwest of the United States are among the largest trees on earth. Moving south in the Northern Hemisphere, we come to the temperate broad-leaved and mixed deciduous/evergreen forests that once dominated the landscapes of Europe but are now largely gone, replaced by farmlands, cities, towns, and roads. There are still large areas of this type of forest in the United States. And so on down to the tropics of South America, Africa, and Asia, to the luxuriant rainforests, where the warm, wet, and stable conditions encourage enormous plant and insect biodiversity. There is considerable variation within each of these broad categories, and we can't possibly discuss them in detail, but they differ sufficiently in their diversity and the rates of the processes that determine their responses to the environment to justify treating them as separate entities.

With the exception of Antarctica, where there is virtually no vegetation of any kind, the vast expanses of ocean ensure that temperatures in the Southern Hemisphere are generally much more moderate than in the north. The tropical rainforests that span the globe at the equator reach southward in South America, Indonesia, and the Philippines, but farther south, in Australia, South America, and southern Africa there are large areas where the low rainfall will not support the growth of forests. Forests in Australia range from a small, almost relic strip of tropical rainforests in northeast Queensland to broad-leaved evergreen—largely eucalypt—forests along the east coast, and in the southwest corner of Western Australia. There are large areas of temperate rainforest, tall eucalypt, and mixed forests in Tasmania, and a thin belt of temperate forests in the southwestern foothills of the Andes, in Chile. South Africa has woodlands rather than closed forests, of which there are some small areas in sheltered locations, mainly on its eastern side. The South African forestry industry is dominated by plantations.

Climate might be called long-term weather. It describes the average conditions over seasons and years. The climate of land areas is determined by the size of the land mass, latitude, topography, and distance from the oceans. Interacting with soil type and soil characteristics, climate determines what kind of trees can grow in any particular area, and how fast they are likely to grow. To understand how that happens, we need some understanding of the biophysical processes involved. We also need this background because most management actions imposed on forests affect the way trees interact with the weather. These are outlined in chapter 3, which is, in essence, a short course in forest ecophysiology; it is intended to supply the framework essential to our objective of identifying problems and proposing possible solutions. (We have provided much more detail in textbooks: see Waring and Running 1998, 2007; Landsberg and Gower 1997; Landsberg and Sands 2010.)

Trees grow because they capture CO_2 from the air and, through the process of *photosynthesis*, convert it into carbohydrates, using the energy of the sun. Photosynthesis is the fundamental biological process, the one that drives most life on earth. For it to work, trees (and all other land plants) need water, favorable temperatures, and mineral nutrients. Photosynthesis is driven by light—short-wave radiant energy from the sun. The process converts CO_2, absorbed through the leaves, into carbohydrates, which have to be moved to the various parts of trees. All of this is what constitutes tree growth.

Because of the immense importance of forests in relation to water, we have placed particular emphasis on the hydrological cycle in chapter 3. Some of the precipitation falling on forests is intercepted and evaporated from the foliage. Some falls through the canopy to be stored temporarily as snow on the surface, or as liquid in the soil where it is taken up by roots, or continues downward to feed streams and underground aquifers. Storage of water in the soil reduces surface runoff with the result that peak flows of streams are reduced and the periods of low flow lengthened as the stored water seeps into the drainage channels. Water stored in the soil is taken up by roots and conducted through the trees to the foliage, where it is returned to the atmosphere as water vapor, by transpiration. Without forests, the flows in many rivers will become far more erratic, with higher peaks and longer periods of low flow.

Forests exist today in a world dominated, and irreversibly changed, by humans. Among the many impacts stemming from that dominance are the climatic shifts caused by the accumulation of CO_2 in the atmosphere, resulting from the burning of fossil fuels and biomass in the course of land clearing. This alters the global energy balance, which is causing global air temperatures to rise. Methane is an even more potent "greenhouse gas" than CO_2, and its concentrations in the atmosphere are also rising as a result of increasing livestock numbers, wetland drainage, landfills, and melting permafrost. The rising temperatures caused by these changes to the atmosphere are having direct effects on ecosystems— including forests—and on human societies, and these are likely to become more marked in the foreseeable future. We have therefore provided in chapter 4 a brief treatment of the physics underlying global climate change, some consideration of its likely magnitude, and discussion about the probable effects on forests and forest growth. Although scientific opinion about it is almost unanimous, we acknowledge that there are significant numbers of people who deny the anthropogenic basis of global warming, and even assert it isn't happening, so we assess their arguments.

We also consider some of the direct and indirect effects of climate change on forests, in terms of drought, insect attacks on stressed trees, and the increased frequency and intensity of wildfires that can be attributed to climate change.

We have said that the forests are intimately associated with the history of humanity. From time immemorial we have lived in them, used their products, and, as we have seen, destroyed them. And we still do. In chapter 5 we consider in some detail how we value and use forests and forest products in the modern world. Wood products are immensely important while the less well-recognized ecosystem services that forests provide—carbon storage, reliable supplies of clean water, and biodiversity—are of universal value. Despite the predictions of paperless offices made when computers began to dominate our lives, it hasn't happened: the world's consumption of paper has continued to grow so that demand for the material from which it is made—wood pulp—has become even greater. We should note, of course, that statements of this sort have to be considered with some care: the demand for any commodity depends on per capita demand, and on the number of people. So per capita demand may be stable or even going down, but human populations continue to rise, so overall demand for most commodities tends to go up. This applies to other products of the forests, such as wood for furniture and building (and for chopsticks).

In developed countries wood is now used much less as a fuel than in the past. As we pointed out earlier, in our brief survey of historical deforestation, the forests could not supply the voracious demands of industrial processes for fuel—even industrial processes several thousand years ago. But in poor countries, across Africa particularly, and in Asia, wood or charcoal for household fuel is a constant problem; finding enough of it to "keep the home fires burning" can involve long walks searching for firewood, which then has to be carried back (usually by women). For this reason the establishment and management of plantations of suitable, fast-growing species is an important aspect of development and aid. But collecting firewood is a minor cause of deforestation. We have already noted, earlier in this chapter, the problem of tropical deforestation. We consider this in more detail in chapter 5.

Chapter 6 deals with forest management. We provide some background on the basic principles of conventional economics, noting the differences in approach that arise from different ownership of the land, and the difficulty of applying (neo) classical economic theory to forestry. Among the shortcomings of conventional economics is the fact that most countries assess their economic growth in terms of gross domestic product (GDP), but none factor into its calculation the natural capital that forests (and indeed all ecosystems) represent. We consider this in chapter 6 and go on to discuss forest management policies in different countries and the economics of forest operations on public and private land. Some simple economic modeling illustrates the basis of decisions that the managers of plantation estates have to make. The probable yield of wood from any area is estimated using growth and yield models. These are commonly empirical, but the use of process-based models, and of remote sensing to measure the characteristics and estimate the growth of forests, is becoming more widespread. These issues are revisited in the final chapter.

Fire is an inescapable fact of life for forests, and it's not always a bad thing. Some forests are adapted to fire, and in fact need it to keep themselves in a healthy state. Australia probably has the best examples of these, and fire is an important management tool in many Australian forests. But even fire-adapted forests are susceptible to fire in some circumstances—such as during unusually severe droughts when high-intensity fires may occur. Whether some trees survive those fires or all are effectively destroyed depends on a number of factors and circumstances. In the last part of chapter 6 we expand on the subject of fire management in forests.

There are virtually no pristine forests in the world today, and the rates of forest loss around the world are alarmingly high. In chapter 7, we outline the management practices that will be needed around the world to ensure the survival of the large areas of beautiful, valuable forests that still survive in various locations. One of the keys to their long-term survival must be a continual increase in the amount of wood coming from plantations, although new plantations should not be established at the expense of native forests. Large areas of single-species forest are not resilient systems; active steps have to be taken to ensure their resilience, and that the production from them is sustainable.

A Look into the Future

Any prediction about the future of forests must take account of the fact that all natural systems are under enormous pressure because of the material success and overwhelming fecundity of the human race, and the related impacts on the earth's climate. Since our book does not pretend to present comprehensive data, or be a comprehensive survey of the current situation with regard to forests, our predictions are based on our own experience, observations, reading, discussions with professional colleagues, and many years of visiting forests around the world, in the company of local experts and managers.

There is a realistic chance, in developed countries, of preserving areas of forest large enough to provide not only the economic products that societies need from them, but also the recreational and aesthetic opportunities associated with forests. The legal and political structures, and the traditions of conforming to established legal systems in such countries, provide some guarantee that public order will not break down. Where it does break down forests are likely to be degraded or destroyed by people looking for fuel, shelter, or space to plant food crops. This is already happening in poor countries across Africa and parts of Asia, where human population growth is out of control and the opportunities for some form of income are scarce. In many of these countries the pressures on forests are such that the probability that large areas will survive must be low. These pressures derive not only from the historical need of the poor to find wood for fuel and construction, but also arise because of widespread corruption. In many of the countries with significant areas of tropical rainforest, powerful individuals in government frequently issue licenses to commercial interests, both domestic and foreign, to exploit the forests for valuable hardwoods, without safeguards or oversight. The result is destruction of large areas.

So where do we go from here? We have argued that it is important to understand the ecology of forests, how they grow, how they are likely to respond to exploitation of various sorts, and the pressures imposed by humans. We recognize that, in the final analysis, the decisions made by societies, which will include the decisions about protecting and managing forests, are determined more often by politics and economics than by scientific considerations. Science cannot provide answers to political questions, although it should be an important factor in economic analyses. It is our view that decisions will be better if their implications are assessed in the light of scientific knowledge and understanding. So we are prepared to say (chapter 7) how we think the (relatively near) future might look in relation to forests, and to indicate the sort of policies and decisions that, we believe, will lead to the best results for human societies and the beleaguered forest ecosystems of our crowded world.

Chapter 2

Forest Types around the World

In this chapter, we provide an outline of the principal forest types that occur in the main climatic regions of the world, to gain an appreciation to how each is adapted to its own environment. Since the geographical distribution of natural forest types is largely determined by climate, we also provide some information about the climates of the main forest areas.

The area of the globe covered by forests has been and, as we noted in chapter 1, is being rapidly reduced by humans. This contraction of forests began with the development of agriculture but has accelerated in recent times: during the last three hundred years, according to the Millennium Ecosystems Assessment (2005), the area of forests has been reduced by half. Forests have effectively disappeared from twenty-five countries and another twenty-nine have lost about 90 percent of their forest cover.

Although the processes that govern the way trees grow are basically the same for all species, as we shall see in chapter 3, the forest types that occur in particular areas are different because plants have developed, through evolution, all sorts of adaptations that allow them to grow, survive, and reproduce in varying environments. But the world is now experiencing rapid changes in climate (discussed in chapter 4) that may alter the frequency, intensity, and kind of disturbances that forests evolved with, and are adapted to, leading to changes in forest type. On a different time scale, occasional catastrophic events that kill numerous trees may be fatal to some forest types, but lead to regeneration of others. It's important to recognize the consequence of different types of disturbance and variation in their frequencies.

We tend to characterize forest types by assuming that they still exist in their pristine state, although very few do today. Humans have disturbed most of them to some degree and have also been instrumental in the movement of tree species from one continent to another, allowing some to naturalize in their new environments. Among the most important examples are the commercial plantations of exotic tree species grown in short rotations, as crops. These are replanted after harvest and attempts are made to improve the growth rates of succeeding plantings through genetic and cultural modifications. Species diversity in commercial plantations is deliberately restricted, but their high yields per unit of land help to forestall the loss of biologically richer native forests in many countries.

Main Forest Types

In many books on forests, detailed descriptions of the species present in different habitats are provided, but in this one we treat forest types in broad general terms, on the basis of their distribution by latitude from the most northern (boreal) through the temperate, to subtropical and tropical zones. Obviously the regions outlined in figure 2.1, designated boreal, temperate coniferous, and so on, are not completely covered in forests of those types; they all include areas altered by humans—via the establishment of towns, roads, and cultivated fields. Also, the areas of natural forest that do exist within these broad categories are seldom spatially homogenous. They include patches, of varying size, which provide habitats for species that

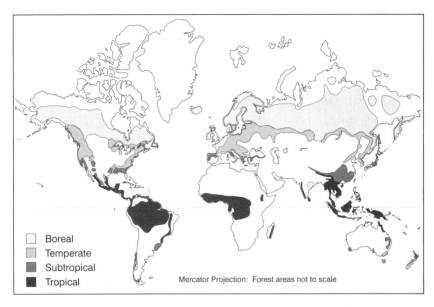

FIGURE 2.1. Map showing the distribution of the main forest types around the world. Since the map is a Mercator projection, land areas are distorted, particularly toward the poles, and should not be taken to indicate relative areas of different forest types. However, it shows their distribution and illustrates the fact that there is far more forest in the Northern than the Southern Hemisphere.

may be different from those in surrounding areas. These differences may be the result of small differences in soils and topography, such as gullies or hillsides. Differences in disturbance history, such as the time since the last fire, can also result in significant changes in forest types within an apparently fairly homogenous region, for example, because more recently burnt areas are at different stages of succession. Nevertheless, the broad categories indicate the forest types that we are most likely to find in each region.

On Being Evergreen or Deciduous

It's worth digressing, briefly, to consider some of the more important characteristics of evergreen and deciduous trees.

The defining characteristic of evergreen trees is that, under normal circumstances, they retain some foliage through all seasons.

The length of time that a particular generation of foliage is retained tends to be characteristic of a species, but is also affected by environmental conditions. Periods of water stress or damaging frost, for example, result in premature shedding of foliage. On the other hand, long-lived trees, growing slowly in cool, moist environments may hold some of their foliage for more than thirty years.

There are two main types of evergreen trees: those with needle-like or scale-type leaves and those with broad leaves. The first type, with few exceptions, is represented by conifers related to the more ancient family of Gymnosperms, all of which bear naked seeds, usually held within cones. Conifer foliage usually persists on the trees for three to five years and, in a few species, for a decade or even more. In contrast, Angiosperms, representing the great family of flowering plants, have broad leaves which, even when the trees are evergreen, generally last only a year or so—sometimes for only a few months—before being replaced.

In warm, dry climates, deciduous trees lose their foliage seasonally in response to drought, whereas in cold climates foliage loss is associated with shortening hours of daylight in the autumn. In cold-climate species the reduction in daylight hours results in superb displays of color as green pigments are broken down to display yellows and reds that absorb and reflect light in different wavelengths. The breakdown of pigments is coupled with the mobilization of nutrients that are translocated from the dying leaves for storage in twigs and other living cells of the trees.

Boreal Forests

Boreal forests occur exclusively in the Northern Hemisphere, in a band 400 to 600 km wide stretching from Scandinavia to eastern Siberia, through Alaska and across northern Canada. Today, they comprise about a third of the total area of forested land in the world, which Waring and Running (2007) estimated has been reduced by humans by about 20 percent over the last few hundred years.

The climate of the boreal regions is among the harshest in which

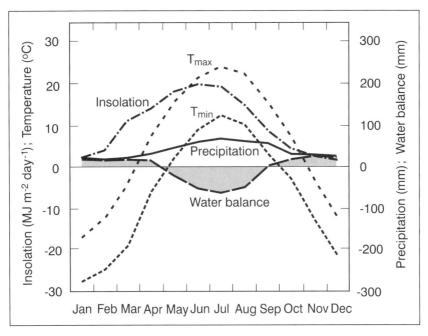

FIGURE. 2.2. Climate diagram for a boreal region. The curves reflect average monthly values of incoming short-wave radiation (insolation, MJ m^{-2} day^{1}), maximum (T$_{max}$) and minimum (T$_{min}$) temperatures (°C), precipitation (mm rain or rain equivalent, as snow), and the water balance, calculated as the difference between precipitation and evapotranspiration (see the section on hydrology in chapter 3). Redrawn from Landsberg and Gower (1997).

trees can survive. Winter temperatures usually fall to −30°C and may fall as low as −50°C. Figure 2.2 is a fairly typical climate diagram for a boreal region. It shows that the amount of solar energy available for tree growth is generally low, except during long summer days when the total amount received is as high as in much lower latitudes (the data in the diagram are for 53°N; compare with figure 3.3). Annual precipitation also tends to be low. The growing season for boreal trees is only four to five months, during which air temperatures may reach a maximum of 15°C to 20°C, with daily averages of 8°C to 10°C. The low temperatures—particularly at night—mean the air cannot hold much water vapor, and it tends to be very dry. A series of measurements of atmospheric humidity in this region showed that, for most of the year, the atmospheric vapor pressure deficit, which drives transpiration (see chapter 3), was around 1 kPa (equivalent to 10 millibars of mercury), rising to about 2 kPa in August.

Boreal forests tend to contain relatively few tree species, many of which are in nearly pure stands that establish themselves after major disturbances, to be displaced by more shade tolerant species that grow up through the canopy. Stand structure tends to be simple. The conifers, in particular, are likely to have narrow crowns and flexible branches, favoring light absorption at low sun angles and efficient shedding of snow. Boreal species, as well as those adapted to high-elevation areas in mountains farther south, have evolved a number of mechanisms that allow them to tolerate very low temperatures. They have the ability to expel free water from living tissue during the autumn and to bind the remaining water to cell walls. When fully cold-hardened, which happens when they have been subjected to progressively falling temperatures, they can even withstand being frozen in liquid nitrogen at −196°C. (This has been done in experiments: −196°C is a long way below any temperature that would be experienced in nature.) In addition, root membranes can efficiently absorb water in the spring at only a few degrees above freezing. Most of the tree roots are restricted to depths of 20–30 cm because ice often remains present below that level, or the substrate is waterlogged. Frost can occur at any time during the summer, but the buds and new foliage are rarely injured unless temperatures fall below −5°C.

Pines (*Pinus*) and spruces (*Picea*) are the most common evergreen species in the boreal regions. Deciduous trees include larches (*Larix*) as well as a few Angiosperms such as birch (*Betula*), aspen (*Populus*), and willows (*Salix*). None of these species depend on birds or insects to distribute their pollen; that's done by wind. Lichens, a synergistic (mutually beneficial) alliance of a fungus and bacteria, populate the understory in high-latitude upland forests and serve as a critical food source for caribou or reindeer. There is also a wide range of berry species. On wetter ground, a series of moss species align along moisture gradients. Although the natural productivity of boreal forests is generally low, this is not because species lack the capacity to grow faster but because the process of decomposition is very slow so that nutrients, particularly nitrogen, become very slowly available to trees. Adding enough fertilizer can increase tree growth rates three- to fourfold (Bergh et al. 1999).

The native flora and fauna of boreal forests are still largely intact, because few invasive species are adapted to such harsh conditions. The major disturbances, which the forests have evolved to cope with, are wildfires and outbreaks of native insects, but the intervals between these disturbances are becoming shorter, and their intensity greater, as the climate warms (chapter 4). Because boreal areas are sparsely populated by humans, and the commercial value of trees per unit of land is low, the ability to control outbreaks of fire, insects, and pathogens is limited.

Temperate Forests

The forests in the cool temperate regions of the world vary from predominantly coniferous to predominantly broad-leaved hardwoods. All of them are mixed, to various degrees, but we have chosen to deal with three major categories: temperate coniferous, mixed temperate, and broad-leaved evergreen.

Temperate Coniferous Forests

Coniferous forests in the temperate regions represent only a small percentage of the world's forests but contain some of the oldest and most massive living trees on earth. These forests are heavily exploited for wood production because the trees are generally straight and tall and contain wood easily milled and dried for commercial use.

Conifers are the dominant trees in mountainous areas of North America, Europe, and China; smaller areas of temperate conifers also occur in mountainous regions of Korea, Japan, Mexico, Nicaragua, and Guatemala. *Pinus* species occur quite widely in the Southern Hemisphere. Species diversity in temperate coniferous forests is much greater than in boreal forests, but they contain fewer species per unit land area than tropical rainforests.

In the coast redwood (*Sequoia sempervirens*) forests in northern California trees may reach heights of 100 m and accumulate standing biomass of more than 3,000 t ha^{-1}, while stands of Douglas fir (*Pseudotsuga menziesii*) in the Pacific Northwest of the United States and Canada may exceed heights of 70 m with 1,000 t ha^{-1} of standing biomass, plus an additional 300–500 t ha^{-1} of standing dead and fallen trees (Waring and Franklin 1979). The only comparable biomass figures to these are eucalyptus forests growing in the wet, temperate climate in southeastern Australia.

In areas where temperate conifers are the dominant species, summers tend to be warm and winters cool and wet. Precipitation varies widely, ranging from as low as 500 to more than 2,500 mm yr^{-1}. The data in figure 2.3 are for Seattle (latitude 47°N), on the northwest coast of the United States; the temperature pattern is fairly typical of the region but warmer in the winter and drier in the summer than many regions where temperate evergreen forests occur. In the Pacific Northwest these conditions favor the dominance of temperate evergreens. Elsewhere, temperate evergreens are more likely to colonize recently disturbed areas where, if the sites are not disturbed again for many years, the evergreens are eventually replaced by deciduous broadleaf species. Most tree species in temperate forests can tolerate subzero temperatures, but not the extremes of the boreal regions. In regions with climates like that in figure 2.3, the relatively high summer temperatures and solar radiation are likely to result in high transpiration rates. If precipitation falls below transpiration for long periods, causing the negative water balance in summer, as shown in the diagram, the trees will be subject to severe moisture stress unless enough water is stored in the soils to make up for the deficit.

Many temperate conifers are well adapted to dry and infertile conditions: they tend to occur on soils that cannot supply water and nutrients in the amounts needed by deciduous species. The waxy foliage surfaces and sunken *stomata* (the tiny pores on leaf surfaces through which water vapor is lost and CO_2 is absorbed

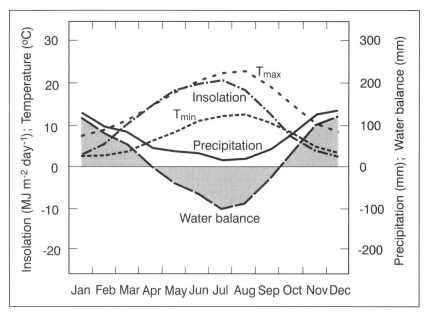

FIGURE 2.3. Climate diagram for a region with temperate coniferous forests—in this case, Seattle in Washington State, USA. The curves describe the same variables as those in figure 2.2. Redrawn from Landsberg and Gower (1997).

from the atmosphere [see chapter 3]) help reduce water losses. A large proportion of their stems consists of sapwood that serves as a storage reservoir, so the trees can tolerate periods of drought that would damage many broad-leaved species. Common genera in the temperate coniferous forests include pine (*Pinus*), fir (*Abies*), spruce (*Picca*), hemlock (*Tsuga*), and cedar (*Thuja*).

Nutrient concentrations in the needles of temperate conifers may be only half those in deciduous hardwoods. These lower concentrations translate into lower demand for nutrients, yet the long-lived foliage provides a greater store of nutrients, which may be transferred into other parts of the trees during peak growth periods, than is available in deciduous hardwoods. Evergreen conifers, with foliage present at all times, can also take up nutrients from cold soils, as long as the temperature remains above freezing, whereas deciduous trees can only take up nutrients when foliage is present (Waring and Franklin 1979).

Mixed Temperate Forests

Mixed forests, containing deciduous and evergreen species, occur throughout the world's temperate regions, particularly the southeastern United States, Europe through northern Iraq and Iran, and China—where the climates are similar to those described for temperate coniferous forests. Most temperate deciduous forests occur in the United States, eastern Asia, and Europe. They contain a wide variety of species, which usually include oaks (*Quercus*), maples (*Acer*), beeches (*Fagus*), and hickories (*Carya*). In more Mediterranean climates, with cool, damp winters and hot, dry summers, some of the broad-leaved species are evergreen with tough, leather-like (sclerophyllous) leaves adapted to long periods of summer drought.

All forests are dynamic, that is, their composition, in terms of species dominance, canopy structure, and the distribution of tree sizes, changes as trees die, and younger ones push through the canopy. Such changes are likely to be more rapid in mixed forests subject to relatively frequent disturbance, which is the case with almost all of the temperate mixed forests. Introduced pathogens such as Dutch elm (*Ulmus*) disease and chestnut (*Castanea*) blight have selectively removed major species from the landscape in Europe and America. Many temperate forests have been selectively harvested or converted completely to other land uses, so it's almost impossible to make general statements about the likely composition of forests in any particular area. This will depend on the history of disturbance and on factors such as the seed bank available for regeneration and the weather patterns that followed a major disturbance. The composition of a regenerating forest may not be the same as that of its predecessor. Comparing climatic conditions with the current distribution of mixed temperate forests determined from satellite imagery, Waring and Running (2007) estimated that, worldwide, the area of mixed temperate forests has been reduced by 40 percent from the original area.

Broad-Leaved Evergreen

The remarks in the previous paragraph about disturbance and the consequent uncertainties in composition also apply to temperate broad-leaved evergreen forests in Japan, Chile, New Zealand, and in scattered remnant patches in Asia. Temperate broad-leaved evergreen forests in South America include a range of types that occur from lowlands to the slopes of the Andes. They have all been heavily exploited for timber for construction, and many of these forests have been destroyed or are seriously degraded. Large areas have also been cleared for agriculture. Timber extraction continues today.

The forests of Australia represent the greatest continual area of sclerophyll forest still in existence. There are more than six hundred eucalyptus species distributed across Australia in a wide range of habitats, although only a few (notably *Eucalyptus pauciflora*) tolerate the cold and snowy conditions in the mountains of New South Wales and Victoria. Eucalypts are also scarce in Tasmanian rainforests and in the arid interior of the continent, although they can be found there along streams and in sheltered areas in isolated ranges.

We said earlier that natural forests are seldom spatially homogenous: large variations in forest type occur in any region where there are hills or mountains, rivers, and streams. Soil variation is always an important factor, but so too is receipt of solar energy: in the Southern Hemisphere north-facing slopes receive more energy, and so tend to be hotter and drier than south-facing slopes. The reverse situation applies in the Northern Hemisphere. We discuss solar (radiant) energy in more detail in the next chapter.

An interesting example of the influence of small changes in physiography can be found in the Blue Mountains of Australia. The Blue Mountains are essentially layered sandstone, deposited in past geological ages by huge rivers flowing from the west. The soils derived from the sandstone are generally shallow, poor in nutrients, with low water-holding capacity, so that despite relatively high average rainfall (about 1,200 mm per year) the vegetation is mainly sparse shrubs and low (4–5 m) eucalypt woodland. However, there are taller forests where the soils are deeper. Small areas of basalt-derived soils, from volcanic caps that were forced up through the sandstone, provide patches of deep, fertile soil. But the vegetation is also influenced by aspect. In the area of Mount Wilson, where the soil is derived from basalt, there is an east-west ridge, the north-facing slope of which carries open, tall eucalypt forest (*E. viminalis* and *E. fastigata* are important species). Crossing to the south-facing side, the ridge falls away to a stream running eastward and the vegetation changes abruptly to cool temperate rainforest, dominated by sassafras (*Doryphora sassafras*) and coachwood (*Ceratopetalum apetalum*), with an understory of ferns and tree ferns (*Cyathea australis*). The big differences in vegetation type are caused by the fact that *insolation* on the southern slopes is much lower than on the north-facing slope, so the north-facing slope is drier.

Since colonization of Australia by Europeans, almost two-thirds of the natural forests that were present in the eighteenth century have been destroyed; forests now cover only about 20 percent of the total land area. Their destruction was the result of clearing for agriculture, repeated severe wildfires, and logging. Logging, even using the clearcutting[1] procedures that have been practiced, particularly since the advent of heavy machinery, does not necessarily destroy forests, but it does tend to change their species composition and structure, particularly when the logging debris (foliage, bark, branches, the top parts of the trees) is left to dry on-site and then burned. The forest that regenerates after that treatment depends on the seed bank that was in the soil, on weather conditions, and on the presence of invasive species.

All eucalyptus species are evergreen hardwoods. They may shed leaves at any time of the year, and are particularly likely to do so during periods of water shortage. The *leaf area index* (LAI) of the trees (total projected surface area of leaves per unit of ground area) tends to be low, usually below 3 (compare this with the values of 8–10 often reached by the tall coniferous forests in North America), so in most Australian forests there is enough light for dense understory to develop. The understory may be so dense that it inhibits the growth of eucalypt seedlings. Canopies of the majestic *Eucalyptus regnans* in Tasmania and Victoria—the tallest flowering plants in the world—may reach 100 m in height and accumulate biomass rivaling the redwood and Douglas fir forests of the Pacific Northwest in North America. The climate in these areas is moderate, with warm summers and mild, although not particularly warm, winters: winter maximum temperatures are in the range 10°C–15°C and mean minimum of 2°C–3°C. Snow is common in winter, particularly in the higher areas. Precipitation, ranging from 1,400 to 2,000 mm per year, occurs all the year round, but is generally heaviest, and more reliable, in winter. Summer temperatures may reach more than 30°C; infrequently they may go much higher than this, usually in dry periods. The risk of fire is very high at such times.

Eucalyptus forests extend up the east coast of Australia, and also occur in the southwest corner of Western Australia, where two species, Jarrah (*E. marginata*) and, on the better soils, Karri (*E. diversicolor*), have been the basis of the forestry industry in that state. Jarrah is a slow-growing tree that produces a superb red, very hard timber. Karri grows in more favorable areas and produces a more typical eucalypt timber. The most common and widely distributed of the eucalypts is *E. camaldulensis*, known as River Red Gum. It occurs throughout most of the Murray-Darling basin, particularly in the floodplains, and in central Australia along watercourses.

Most eucalyptus species are well adapted to fire, as are the nitrogen-fixing species of *Acacia* that thrive following fire. But Australia is subject to recurring droughts, which, when combined with hot weather and windy conditions, periodically produce wildfires of frightening intensity. The native people of Australia used fire for thousands of years in a manner that appeared to be consistent with natural regimes: fires caused by lightning strikes. Their practices, and the relatively frequent natural fires, resulted in usually low-intensity fires that did not kill mature trees. Since European settlement, forest management has been greatly changed: modern human communities are ill-adapted to fire, so it has been excluded as far as possible from the remaining forests. The result has been fuel buildups so that when fires inevitably do occur they are likely to be much more intense, damaging to the forests, and dangerous to humans and their dwellings than was normal historically. We discuss fire regimes in forests, the management of fire and fire as a management tool, in some detail in chapter 6.

At various locations in the Southern Hemisphere there are relatively small areas of evergreen *Nothofagus* ("beech") forests, notably in New Zealand, Tasmania, and Chile. *Nothofagus* is particularly interesting because southern beech forests evolved some 100 million years ago, when Australia was still part of the supercontinent *Gondwana*. Between 160 and 65 million years ago, Gondwana split apart, forming today's southern land masses: Antarctica, South America, Africa, Madagascar, Australia, New Guinea, New Caledonia, and New Zealand. The relic old-growth *Nothofagus* forests, particularly in New Zealand and Tasmania, are spectacular examples of primeval forest. Despite being on different continents for thousands of years these trees are still classed as one family (Nothofagaceae).[2]

Tropical Forests

Tropical evergreen forests—generally called rainforests—comprise ~ 50 percent of the world's forests. According to Waring and Running's (2007) comparison between satellite surveys and environmental correlations, about 25 percent of the original forest has been converted to agricultural or related use, although this proportion has probably recently risen substantially. Tropical evergreen forests occur in areas where temperatures are relatively high with little seasonal or diurnal fluctuation. The greatest single area of these forests is the Amazon basin in the northern half of South America, followed by the forests of central Africa (Congo region, Gabon, Cameroon, and southeast parts of Nigeria) and the forests of Southeast Asia. All these regions are characterized by more or less constant mean daily temperatures in the mid to high 20s (°C), with little difference

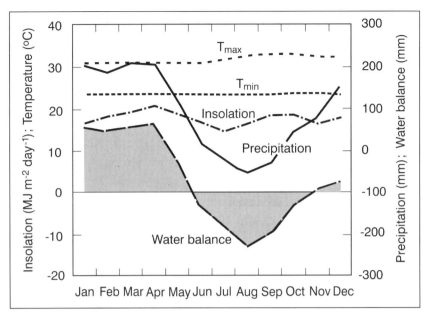

FIGURE 2.4. Climate diagram for a region of tropical forests—in this case Manaus, on the Amazon River in Brazil. The curves describe the same variables as those in figure 2.2. Note that there is a quite extended dry period (negative water balance). Redrawn from Landsberg and Gower (1997).

between average maximums and minimums. Since they are located along the equator, radiant energy income is fairly constant through the year, although it varies slightly, depending on cloudiness. Contrary to general belief, not all tropical forests are wet all of the time; there are large differences in the rainfall patterns, with some areas experiencing a significant dry season (see figure 2.4, where the data are from Manaus [latitude 3°07' S], on the Brazilian Amazon). In such areas, deeply rooted species (some to 10 m) have an obvious advantage over more shallowly rooted species. The large differences in climate and parent material have a pronounced impact on the structure and function of tropical forests.

Tropical evergreen forests are the most diverse terrestrial ecosystems on earth, with the largest number of species per unit area (frequently more than 300 tree species per hectare; a maximum of 473 per hectare was recorded in the Ecuadorian Amazon [see Waring and Running (2007), 310]). The Amazon forests contain more than 2,500 different tree species with thousands more in the African and Asian forests. African rainforests seem to be relatively poor in species compared with those of America and Asia.

The diversity of species in a wet tropical forest varies significantly across relatively small distances—even from one side of a major river to the other. Scientists who have studied the evolution of these forests believe the reason that localized variation is so high is in part a response to the selection pressures generated by year-round predation by insects and infection by disease. Forests composed of only a few species would be vulnerable to devastation. With hundreds of tree species present, the danger of extinction is much reduced, but there is a tradeoff: trees must produce an array of flowers with different forms and attractive odors to entice specialized insects, birds, and bats to pollinate their flowers. They must also produce specialized fruits, nuts, and berries of various kinds that can be widely distributed and made viable for germination by passing through the digestive tract of different-sized animals.

Another possible reason for the high species diversity is the fact that thousands of years ago, when the sea level was one hundred meters higher, large islands formed that allowed evolution to proceed in comparative isolation (Hall and Harvey 2002). The large river basins of today are a legacy of that period. Therefore populations, particularly plant populations, are confined to relatively restricted areas within which there are also instances of coevolution between plants and insects or animals. Over long periods, all this has led to immense localized biodiversity,[3] rivaled only by that of a few coral reefs.

Tropical forest soils are generally very poor and inherently infertile: high rainfall over thousands of years has leached out mineral nutrients. They sustain luxuriant forests because of the high turnover rates resulting from rapid decay of the organic matter constantly falling from the canopies. The process releases nutrients that are almost immediately reused. When the forest is removed, nutrients are leached from the soils, and the soil structure, normally maintained by tree roots and the activity of microorganisms, breaks down. But there are some soils, produced by indigenous people 500 to 2,500 years ago, which contain large amounts of organic carbon and supported continuous agriculture. These fertile soils were created by incorporating incompletely combusted biomass carbon (charcoal) into the soil, along with massive amounts of green vegetation. This kind of *bio-char* management is recommended as a way to reduce the rate of tropical forest clearing associated with *slash and burn agriculture.*

Rainforest canopies are characterized by layered architecture with, very generally, an upper layer of tall trees—called emergent because they stick out above the main canopy—a middle layer, and an understory canopy of small trees, shrubs, and bushes of various sorts. The standing aboveground biomass of tropical rainforests obviously varies enormously from place to place and depends on the amount and seasonal distribution of precipitation, as well as the soil type, and time since the last major disturbance. Windstorms play an important role in creating gaps in tropical forest canopies. Within these gaps, seeds of light-demanding, quick-growing species germinate rapidly. Some of these are nitrogen-fixing species, members of the legume (Leguminosae) family. The rate of biomass production (*net primary production*) of tropical rainforests is among the highest of all forests, but the actual annual increase in standing biomass is not particularly high. This is because leaves, twigs, and branches are constantly falling to the ground—the process is called litterfall—where they are consumed by insects and animals and recycled through them. The litter is also rapidly decomposed by microorganisms, releasing the nutrients it contains, which are almost immediately taken up by the living vegetation. Where rainforests are cleared for agriculture this cycle is disrupted, and within a few years the soils will support only poor grasses or shrubs.

Tropical rainforests are under heavy pressure from human activities. There is general consensus, among a variety of sources, from studies of various reliability, that they are being destroyed at the rate of about 14 million hectares per year. It's difficult to get accurate and unequivocal figures, because definitions of what constitutes tropical forest vary, and there is also uncertainty about how much is completely destroyed, that is, damaged beyond the *tipping point,* at which the land could revert to forest (see the section on *resilience* in chapter 1). Waring and Running's (2007) comparison between satellite surveys and environmental correlations indicated that about 25 percent of the world's original tropical forest has been converted to agricultural or related use. This figure is one of the lower estimates of total area destroyed. We discuss the use and abuse of tropical forests in some detail in chapter 5.

Tropical forests provide a prime example of ecosystems that appear to survive and thrive in relatively stable states as long as disturbances, even the extensive blowdown that may result from cyclones, do not completely remove functioning canopies. Disturbances that create large gaps and expose the soil around intact forests can push them across thresholds into alternative, degraded states, from which they cannot recover, even if replanted, if their highly selective native pollinators and seed-dispersing animals are absent. Although diversity among tree species is high in the tropical forests, few are adapted to fill gaps large enough to cause the humidity deficit in them to increase substantially. Such gaps are caused by slash and burn farming and clearing for various reasons, such as the establishment of oil palm plantations. Once the edges of a wet tropical forest are exposed, the understory dries, and fire, a disturbance previously rarely encountered, kills indiscriminately.

We consider in more detail, in chapter 6, how tropical forests should be managed—or perhaps we should say, not managed, but conserved.

Plantations and Managed Forests

Plantations are essentially (usually large) areas planted to single tree species, for the purpose of wood production. A large proportion of the timber products used in the world today, particularly pulp-wood, but increasingly sawn timber products, come from plantations rather than from natural forests. Plantations now represent less than 3 percent of the world's forest cover but produce more than 25 percent of our wood products. They have a number of advantages as wood-producing systems: the objective is to achieve a specified product in the minimum possible time, so management is generally much more intensive than that of natural forests. Genetically improved and uniform seedlings—nowadays frequently clonal material, genetically as well as physically uniform—can be planted on pre-prepared land at optimum spacings. It is economically feasible to use fertilizers to improve soil nutrition and to control weeds (Landsberg and Gower 1997). The area under plantations around the world is given in table 2.1.

A very high proportion of the species used for plantation forestry is exotic to the areas where they are grown. For reasons that are of considerable interest from the ecophysiological point of view, exotics very often grow faster in areas where they are introduced than do species native to those areas. Why they do so is not always clear. In the case of eucalypts, for example, widely grown as plantation species, it has been suggested that one of the reasons is that the defoliating insects that are an important feature of the native habitats of the trees are absent when they are introduced to other parts of the world. No

TABLE 2.1. Area in millions of hectares under plantations around the world in 2005 and projected areas up to 2020

Year	*Tropical*	*Non-tropical*
2005	70	120
2010	108	139
2015	166	161
2020	255	187

Sources: FAO Report. 2010. Pepke, E. "Global Wood Markets: Consumption, Production and Trade." http://www.fao.org/forestry/12711-e94fe2a7dae258fbb8bc48c5cc09b0d8 .pdf; J. R. Boyle, J. K. Winjum, K. Kavanagh, and E. C. Jensen, eds. 1999. *Planted Forests: Contribution to the Quest for Sustainable Societies.* Kluwer Academic Publishers, Dordrecht, Netherlands.

doubt this is a factor, but the same can't be said of all species that do well outside their native environments. Part of the explanation probably also lies in the fact that plantation-grown species are, as we noted, generally managed much more intensively than may be the case in their native habitat. They may also be grown on better soils. Eucalypts are grown, mainly for pulp, in Brazil, Venezuela, Colombia, Chile, and Uruguay, among other South American countries; also in Portugal, South Africa, China, and India. In many of these countries the area of eucalypt plantations is greater than in Australia, their country of origin. Forestry in that country has, for a long time, been focused on the natural forests (this is now, finally, changing) and on softwood plantations, although eucalypts are now quite extensively grown for pulpwood. The growth rates achieved in some of the eucalypt plantations, particularly in South America, but also to a lesser extent in Portugal, are spectacular. It is normal to achieve yields of more than 300 tons dry biomass per hectare in six to seven years in Brazilian plantations; that would require fifteen to twenty years almost anywhere in Australia.

The major softwood species grown in Australia is *Pinus radiata*, also called Monterey pine, since it originated from the Monterey Peninsula in California. This species is also the most important softwood in New Zealand and Chile, where it provides the basis for a successful and highly competitive softwood export industry. It is also extensively grown in South Africa. *Pinus radiata* is a particularly interesting example of the better growth that may be obtained in foreign locations: on the Monterey Peninsula it is a relatively unimpressive, rather scrubby pine, but in some of the places where it is grown as a plantation species, particularly in New Zealand, it achieves growth rates and grows to sizes unknown in its place of origin. In general, when North American species are introduced to New Zealand they consistently outgrow local species.

Other major softwood plantations round the world include large areas of Sitka spruce (*Picea sitchensis*) in Britain. This species is native to the west coast of Canada and the Pacific Northwest of the United States, where the trees grow to be very large. There are extensive plantations of loblolly (*Pinus taeda*) and slash pine (*Pinus elliottii*) in the southern United States, and Douglas fir (*Pseudotsuga menziesii*) in western Canada and the Pacific Northwest in the United States. Douglas fir is also grown as a plantation species in New Zealand and Chile.

The managed forests of Scandinavia, Finland, and Europe, many of which are stands of Scots pine (*Pinus sylvestris*) or Norway spruce (*Picea abies*) and birch, are somewhere between plantations and natural forests. They are not natural stands, since they are harvested and thinned, and have been managed in this way from the early nineteenth century. Sweden and Finland were the first countries where forest conservation laws came into effect, in 1886 and 1903, respectively; today sustainable forestry must not only ensure a reliable yield of timber and the multiple use of forests, but also preserve biological diversity.

In southern Sweden, Finland, and Norway, forests traditionally served as a source of wood for construction, fuel, and the production of numerous useful objects and tools. From the Middle Ages, the forests were exploited for wood tar, pitch, potash, and perhaps most important, charcoal for smelting. The conversion of virgin forests to managed forests in the nineteenth century affected both standing timber volumes and the structure of the forest. Virgin forests were characterized by multiple age and size structures, but these have been largely replaced by uniformly structured, relatively even-aged, young forest stands. Under modern management the trees are seldom allowed to reach maturity, yet the wood stocks in Nordic forests have almost doubled in the last 100 years and continue to grow.[4]

Advantages of Plantations

A major advantage of plantations is the environmental benefit of not destroying, or at least badly damaging, native forests for wood products. As we make clear at several points in this book, our position is that the world's natural forests must be protected and conserved as far as possible because they provide ecological goods and services without which the world will become a progressively less pleasant place for humans to live in. Plantations, as we pointed out earlier, are essentially a form of crop and are therefore different from natural forests. They are, invariably, single-species stands, managed for high wood-production rates and relatively easy access for thinning and harvesting. Wood yields per unit land area and time are generally much higher than in native forests (the Nordic and European forests, as we have just discussed, are exceptions). Data in table 2.2 show that the rota-

TABLE 2.2. Rotation length (years) and productivity (Mean Annual Increment [MAI]) for some of the world's more important plantation species, in a number of countries. The first six entries describe broad-leaved species. The following eight entries describe conifers.

Species	Country	Rotation length	MAI
Eucalyptus	Brazil	7	40
E. grandis	South Africa	8 – 10	20
E. globulus	Chile	10 – 12	20
E. globulus	Portugal	12 – 15	12
Betula (birch)	Sweden	35 – 40	5.5
Betula (birch)	Finland	35 – 40	4
Pinus radiata	Chile	25	22
Pinus radiata	New Zealand	25	22
Pinus spp.	Brazil	15 – 20	16
Pinus elliottii and P. taeda	USA	25	10
Pseudotsuga menziesii (Douglas fir)	Canada, west coast	45	6
Picea abies	Sweden	70 – 80	4.5
Picea glauca	Canada, interior	55	2.5
Picea mariana	Canada, east	90	2

The common names for *Pinus elliottii* and *P. taeda* are slash pine and loblolly, respectively. *Picea* species are spruce.

Note: MAI = $m^3\,ha^{-1}\,yr^{-1}$

tion length of plantations varies from about seven years in Brazil to eighty to ninety years for spruce in Sweden and eastern Canada. The mean annual increments (MAI) obtained with eucalyptus in Brazil and Colombia are among the fastest forest growth rates in the world. Multiplying MAI by the rotation length gives an estimate of total wood production (per hectare) over each rotation. Table 2.2 provides some data on the productivity of various plantation species at different locations around the world. The data in the table are, obviously, average values.

The short rotations characteristic of plantations—at least in warm climates—mean that managers can adapt to variables such as changes in the climate, or the advent of pests or diseases that attack particular species, by planting improved or resistant genetic material at the beginning of each cycle. From a straightforward economic point of view, harvesting plantations is far less costly than harvesting an equivalent mass of timber from native forests.

Plantations can be established on abandoned agricultural lands. This may not be ideal from a production point of view, but has additional benefits: the soil is protected from erosion—the canopies reduce or eliminate the direct impacts of raindrops, which in heavy rain displace soil particles that may then be removed by flowing water. Litterfall (leaves, twigs, etc.) on the surface reduces runoff rates (see "Precipitation and Hydrology," chapter 3) and increases soil organic matter, which helps maintain and restore fertility.

Disadvantages of Plantations

Plantations may provide habitats for various birds and other animals, but fauna populations are generally limited since they normally require the diverse habitats associated with old forests where multiaged trees provide shelter from deep snow or wind, cover from predators, nesting sites, and a wide range of food sources. Plantations generally have few standing or fallen snags, limiting habitat for animals that require these resources. Plantation managers shorten the normal growth cycle following disturbance by controlling unwanted vegetation. This results in limiting the populations of animals that depend on early stages of vegetation. Trees in plantations are usually densely planted to reduce competition from shrubs and grass, to limit the size of branches, and to absorb as much light as possible to maximize photosynthesis. Without gaps in the canopy, it is difficult for many native plants and dependent animals to establish and reproduce.

Without the long rotations (lifecycles) of native tree species, many niches essential to a host of animals are lacking. The normal sequence of disturbance by fire, insects, and disease is disrupted, which can limit biodiversity, particularly for birds and other animals with large home ranges dependent on fallen timber, insects, and disturbance by fire to complete their lifecycles.

It's not unusual for people who live in proximity to large areas of plantations to become very averse to them, leading them to characterize the plantations as, among other things, "biological deserts." This is a fairly common attitude about Sitka spruce in Britain and *Pinus radiata* in Australia. (It may well be the attitude about introduced species in other countries, although we have no evidence—even anecdotal—of that.) It's not an entirely valid epithet, although, for the reasons we have just discussed, it does have some basis in fact. Nevertheless, plantations are immensely important as the means of satisfying voracious human appetites for wood and wood products and easing the pressure on natural forests.

Fertilizers and herbicides require energy to produce, and those used in plantations may enter the groundwater and streams, causing ecological problems. The replacement of grass and shrubs by trees with deep roots and evergreen foliage can greatly increase transpiration and evaporation, so that less water is available for agriculture and aquatic organisms. (We note in chapter 3, in the discussion about forests and hydrology, that the establishment of eucalypt plantations is now restricted in some areas of South Africa because they have altered the local hydrological balance, reducing stream flow and hence the water supplies to villages in the lower parts of the catchments.)

Nonnative trees often escape cultivation and become established in natural areas, where they alter the ecological balance. Monterey pine, eucalyptus, Douglas fir, western hemlock, and bamboo have all done this.

Plantations may exhibit less resilience than natural forest systems containing a range of species. They may be more susceptible to damage by fire than native vegetation, particularly if shrub vegetation dies and becomes potential fuel as the canopy closes. But perhaps a more serious issue is that large, contiguous, monospecific (single-species) areas, may suffer devastating damage if a disease or insect pest appears, either endemic or imported, to which the plantation species is highly susceptible. Because of the lack of genetic variability in the host plants there is nothing to stop the pest proliferating, and the result may be disastrous rates of tree death and loss of production. Many examples of this exist and, if it happens, it's likely to cause changes in management regimes, or force managers to change species.

Summary

We provide, in this chapter, information about forest types, some of their characteristics, and where they occur around the world.

The differences between the forest types in different regions are mainly caused by differences in climate. Evergreen and deciduous trees occur in all climatic regions: evergreens may have needle-like or scale-type leaves, or broad leaves. Most deciduous trees are broad leaved. In warm, dry climates they lose their foliage seasonally in response to drought, whereas foliage loss in cold climates is associated with shortening hours of daylight in the autumn.

Boreal forests, comprising about one-third of the world's forested area, occur exclusively in the Northern Hemisphere, in climates that are among the harshest in which trees can survive. Coniferous forests in the temperate regions represent only a small percentage of the world's forests but contain some of the oldest and most massive living trees on earth, while mixed forests, containing deciduous and evergreen species, occur throughout the world's temperate regions. The eucalyptus forests of Australia—all evergreen hardwoods—represent the greatest continuous area of sclerophyll (hard-leaved) forest still in existence. Although soils in the tropics are generally poor and infertile, tropical evergreen forests are the most diverse terrestrial ecosystems on earth. They comprise about 50 percent of the world's forests but are under heavy pressure from human activities.

Plantations of various species now represent less than 3 percent of the world's forest cover but produce more than 25 percent of our wood products. Wood yields per unit land area and time are generally much higher in plantations than in native forests. Most softwood plantations are pine or spruce species; eucalypti are the dominant hardwood, now grown in many countries around the world. The natural forests of Scandinavia, Finland, and Europe are managed like plantations for high wood-production rates.

Chapter 3

Weather and Climate Determine Forest Growth and Type

We outlined in chapter 2 the main forest types and where they occur around the world. The main factor underlying the distribution of those forest types, and the differences between them, is climate. Our job, as forest ecologists, is to establish what it is about different climates that determines the differences in forests. Also, what is it about different tree species that enables some to grow well in the tropics but not in cooler regions, while species that can survive and grow in the cold, harsh, boreal regions do not do well—or grow at all—in what seem to be the much more favorable tropical and subtropical regions? The answers lie in the interactions between the physical and physiological characteristics of trees, largely determined by their genetics, and the climatic factors that act on those characteristics and so affect the way trees grow. There's also the matter of competition among and between trees and other vegetation types.

Physical characteristics include the structure of the tree and the canopy it forms (spreading or compact and narrow), leaf type (broad leaves or needles), and the water-conducting system, which varies between species. Among physiological characteristics, the biochemistry of photosynthesis and respiration—the fundamental growth processes—is essentially similar in all tree species. They are discussed briefly in the next section. And there are important differences between species in the way carbohydrates are transported from leaves to the rest of the tree, in the ability of tissues to withstand freezing, and in the structure of membranes in the roots that influence the uptake of water and nutrients from the soil.

In order to provide the background for considering the effects of weather and climate on the distribution of forests and forest types, and on forest growth, we need to provide a basic description of tree structure and physiology, and a summary of how trees grow and how weather conditions act on them to determine growth. We also have to consider the components of weather and climate. We said in chapter 1 that climate can be considered long-term weather. The climate of any place or region can be described by seasonal or annual average values of temperature, solar energy income, precipitation, and air humidity. These numbers give us a general indication of the climate of a place, but, since averages conceal variation, they might not convey much information about the weather likely to be experienced at any particular time. For convenience, we will discuss the main weather variables under separate subheadings.

How Trees Grow

Structurally, all trees have one or more main stems, or trunks with branches. Forest species commonly have only one trunk, or bole, but some are more likely than others to be multistemmed. The differences are not important for our purpose here; we will assume a single stem per tree. This is anchored by the immensely important root system, through which water and the nutrients essential for tree growth are absorbed. Branches extend outward from the stems. Progressively smaller branches attached to them carry twigs and foliage. The pattern of branching and the way foliage is attached varies tremendously with species and growing conditions. Some trees have spreading branches with almost horizontally exposed leaves;

others have branches that grow up parallel to, and not far from, the main stem, giving narrow-crowned trees. Branching patterns and crown structure are modified by competition—or the lack of it. Tree crowns in high population stands will always tend to be narrow, and the trees will be relatively tall as a result of competition for light. The crowns of trees grown in the open will always tend to be wide.

Trees (and all other terrestrial plants) grow, as we pointed out in chapter 1, through photosynthesis—the process of converting carbon dioxide (CO_2) absorbed from the air by the leaves, into carbohydrates, using the energy in the visible wavebands of solar radiation (light). Photosynthesis takes place in the leaves, which intercept the radiation, so the way the leaves are exposed to the light is important. The carbohydrates produced are transported through the plants and used to construct new leaves, stems, branches, and roots to replace those that die and fall. The energy needed to carry out these growth processes is obtained from *respiration*— the process by which some of the carbohydrates are broken down biochemically, releasing CO_2 back to the atmosphere. Respiration is a temperature-dependent process. Young, actively growing trees absorb and store more CO_2 than is returned to the atmosphere, but in forests dominated by older trees, the amount of CO_2 absorbed and fixed in living tissues is (approximately) balanced by the amount released back to the atmosphere by plant respiration.

Healthy leaves, containing in their cells the mineral nutrients they need, and with access to good water supplies through the conducting system of the plant, will produce more carbohydrates, for a given amount of energy absorbed, than leaves that are short of some important nutrient(s) or lack water. To absorb CO_2, plants need to expose a wet surface. They do this within the cavities behind the tiny pores—known as *stomata*—on their leaves. Stomatal apertures are controlled by cells, called guard cells, which are flexible and respond to differences in air humidity (see later discussion) and also to light. The structure, size, and number of stomata on leaves vary among species, and with growing conditions, but regardless of these differences the stomata of most plants tend to open in the light and close in darkness. CO_2 that has diffused into stomata is captured on the wet cell surfaces lining the cavities and used for photosynthesis. However, there is a price to pay for wet surfaces exposed to the air: water evaporates from them and has to be continuously replaced. This leads to the process of transpiration, which we will discuss in more detail in the section on hydrology. It's worth noting that the process of diffusion of CO_2 into leaves is affected by temperature, but not in the same way as enzyme-dependent biochemical processes.

The rates of water vapor loss from leaves, and of CO_2 diffusion into them, are affected by stomatal aperture: for maximum rates the stomata must be wide open. The diffusion process is also influenced by atmospheric CO_2 concentrations, which affect the gradient of CO_2 from the air into the leaves, so, in general, higher atmospheric CO_2 concentrations lead to higher rates of photosynthesis. But, like everything to do with ecology, the relationship is not straightforward; the effect tends to be stronger in young than in mature trees and is also dependent on the availability of water and the vital nutrient, nitrogen.

Root systems seem to be determined more by growing conditions than genetics. They consist of large main roots—called coarse roots—smaller secondary roots, and a network of fine roots. It's a reasonable generalization to say that a higher proportion of the carbohydrates fixed by leaves is used to grow roots when water is scarce and nutrients in short supply, than is the case when water and nutrients are in ample supply. The net result is that a tree growing in favorable conditions will have a relatively smaller root system than the same tree growing under unfavorable conditions.

The depth into the soil to which root systems penetrate—the root zone—varies with species and soil type. If there are impermeable layers, or clay layers, or water tables at depths where roots might grow under favorable conditions, these layers will inhibit root growth and reduce the root zone, and so the volume of soil that can be exploited by the trees. In some situations and soil types, long taproots may penetrate the soil to remarkable depths; they have been followed down to 28 m in West Australian forests. But no matter how deep taproots can go, most of the root mass of trees is invariably in about the top half meter of soil, which is generally better aerated than deeper layers and is where most of the nutrients are. Even for very large trees we would seldom consider tree root zones to be more than 2 m in depth, except perhaps in very arid regions. Most of that root mass in those top layers consists of fine roots that permeate the soil in the root zone. The higher the density of the fine root network, the more thoroughly it explores the soil and the shorter the distance that water and nutrients have to move to be absorbed by the roots.

Water is absorbed by roots, and moves up through the conducting system of the tree to replace that evaporated through the stomata in leaf surfaces. The water conducting system of trees consists of fine capillaries—*xylem vessels* in hardwoods, or *tracheids*, with pits on their side walls that open and close, in conifers—within which, under favorable conditions, the water columns are unbroken. These unbroken columns extend up the trunks and proliferate through the branches and twigs into the leaves. So when water is lost from the leaves, it is essentially dragged up through the tree, under tension. The tension is transmitted down to the roots where it sets up suction gradients[1] from the root surfaces to the surrounding soil. Water moves across those gradients into the roots.

If the rate at which water moves into the roots is too slow to meet the rate of loss from the leaves, then the stomata close, or partially close, reducing the loss rate. The rate at which water is lost from leaves—the transpiration rate—is driven by environmental conditions, of which more later. If the soil is dry, water uptake rates may be too slow to meet demand, even if the atmospheric conditions are not conducive to high transpiration rates.

Trees fall into one of two groups in terms of their responses to this problem. In so-called *isohydric*[2] species, stomata close early when water stress starts to develop. This has the advantage that water loss is reduced—hence isohydric—but the disadvantage is that the rate of CO_2 absorption is also reduced, so if the drought goes on for long enough the trees could exhaust their carbon reserves and may die. In so-called *anisohydric* species, the stomata remain partially open, so growth can continue, but the tree will continue to lose water from leaves and possibly other tissues. If rates of water loss continue to exceed supply rates, these trees are likely to suffer cavitation: air bubbles get into some of the xylem vessels and they lose the ability to conduct water. This is damaging, and may not be reversible. If the damage is severe enough the trees will eventually die.

Stomatal control of water loss is a vital fine-tuning mechanism governing tree-water relations. Without this mechanism leaves would desiccate and die whenever supply rates from water uptake by roots fell below loss rates through transpiration, even for short periods. Nevertheless, if water shortages are prolonged, foliage will die—earlier in anisohydric than in isohydric species. Foliage shedding is a common response to drought. Drought that continues to the point that leaves are shed causes major loss of forest productivity; the reduced area of foliage on the trees means that, even when the drought breaks, the amount of CO_2 that can be absorbed for photosynthesis is reduced. The leaves have to be replaced, and this uses carbohydrates that might otherwise have gone to stems. If the drought continues long enough, parts of trees, such as twigs and small branches, die back, and eventually whole trees may die, severely reducing the productivity of the forest.

Weather Factors: Temperature

Summer and winter are, of course, defined by temperature: summers are warm and winters cold, but how warm and how cold varies enormously from place to place. Average temperatures decline with latitude, although the relationship is not straightforward, being complicated by the distribution of land masses across the globe and by altitude. Figure 3.1 shows the annual course of monthly average temperature for three places with very different climates and very different types of forests.

Air temperatures anywhere vary seasonally, and also on much shorter time scales. If we say that the average temperature of a particular month is 20°C, this might have been obtained from a

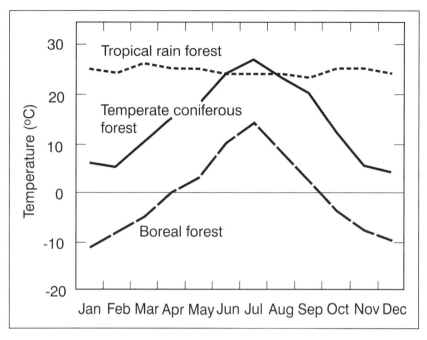

FIGURE 3.1. Annual course of monthly average temperatures in the boreal regions, the coastal Pacific Northwest of the United States (temperate coniferous forest), and the Amazon basin (tropical rainforest). Large seasonal differences in temperatures exert very strong influences on tree growth.

mean minimum (daily) temperature of 7°C and a mean maximum temperature of 33°C. Another month with the same average temperature might have had a mean daily minimum of 15°C and a mean maximum of 25°C. And the mean maximum and minimum temperatures, derived from daily values, may themselves conceal considerable variation. So, within either month it may not be unusual to experience, on a given day, a maximum temperature of 40°C, although that would be more likely if the average maximum temperature was 33°C than if it was 25°C. Annual averages are even less useful in providing information about what can be expected on a particular day. But, notwithstanding all that, we have to use averages to describe climates, otherwise the exercise becomes impossible.

If we want to convey precise information about weather and climate we can provide more statistics. So we might describe the climate of a region as having mean monthly maximum temperatures ranging from 12°C in winter to 31°C in summer, with mean monthly minimum temperatures ranging from –3°C in winter to 11°C in summer. We can add information about the probability of getting specified maximums or minimums in any particular month, the usual number of days with frost, and so on. The expected variation about the mean (lowest and highest totals) is a useful number, usually taken to occur within a thirty-year timeframe.

From the point of view of tree growth, average values provide most of the information we need to explain differences, but we can't ignore short-term fluctuations. A particularly severe frost, or a few extraordinarily hot days, might kill trees in a normally temperate region, so if we're concerned about patterns of tree growth, or why trees die, we might need to look at the probability of the occurrence of extreme events. This includes drought, although that is an event that occurs over a period of weeks to months.

Temperatures are also affected by topography. Air is a fluid and behaves that way. Heat loss from high areas, particularly on clear nights, cools the air there and the cold air may drain down slopes and accumulate in valleys and hollows, causing frost pockets—not good places to plant orchards. In the Northern Hemisphere, south-facing slopes usually receive more sunlight energy (see the next section) and tend to be warmer than north-facing slopes, although there are exceptions. The opposite applies in the Southern Hemisphere.

Trees cannot control their own temperature—only warm-blooded organisms can do that—and, since metabolic rates are temperature dependent, the rates at which trees can grow by converting carbohydrates into leaves, stems, and roots (see next paragraph) are determined by temperature. So is the rate of respiration, mentioned earlier: the very important process by which carbohydrates are broken down to CO_2 while providing energy to build and maintain living tissue in the plant. This is also the process involved in the decomposition of organic materials by microorganisms. If the rates of growth processes are plotted against temperature, the resulting curve is generally parabolic in

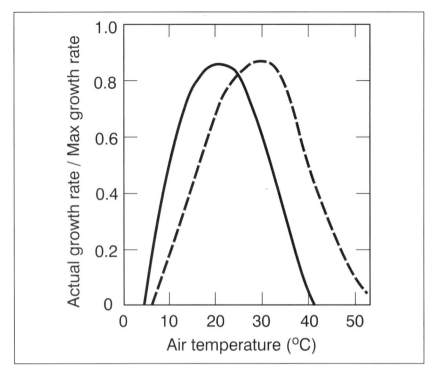

FIGURE 3.2. Temperature response curves for tree growth. The curve that peaks at about 20°C is for conifers adapted to grow best in cool temperate regions, where their growth rates will be highest at about 20°C. The curve that peaks at about 30°C is for tropical species, which would be unlikely to compete with better adapted trees where average maximum temperatures are less than about 25°C.

shape, rising from low rates at low temperatures to the highest rates that can be achieved at some optimum temperature. The rates then fall as temperatures rise above the optimum until, if they get high enough, the enzyme systems that underpin growth can be damaged and the plants may die (see figure 3.2).

The very different optimum temperatures for growth for trees adapted to different environments reflect differences in their enzyme systems; this is the fundamental reason why trees vary in their ability to grow in particular climates. There are other factors: for example, trees such as those that grow in boreal regions and can survive severe frosts have cellular structures and chemistry such that the cells are not damaged by very low temperatures. Their leaves (generally needles) are small and tough. Trees that grow in warm areas with good water supplies tend to have large, soft leaves that are easily damaged by desiccation.

Weather Factors: Solar Radiation

The energy received from the sun as solar radiation varies from place to place and is a major factor determining tree growth. For various reasons, some historical, some relating to ease of measurement, it is rarely considered in normal forecasts and weather information, except that forecasts might say the day will be "sunny with scattered clouds." (Solar radiation and sunshine are not completely synonymous, but for purposes of general discussion they can be taken as the same. There is a close relationship between hours of bright sunshine and the amount of energy [*insolation*] received at any location.)

The daily maximum amount of solar energy received at the surface of the earth at any location depends on latitude and season, which determine the angles at which the sun's rays strike the earth. It's a matter of rather complex three-dimensional geometry, but the equations necessary to work it all out are well documented (see appendix 1 in Landsberg and Sands 2010). Figure 3.3 is a graph of the mean monthly values of *short-wave solar radiation*, that is, energy income at ground level—also called insolation—for a range of latitudes in the Northern Hemisphere. The graph for the Southern Hemisphere is the mirror image, with lowest values in June (winter) and highest values in December. The values on the vertical axis are conventionally expressed in megajoules per square meter of horizontal surface; the scientific shorthand for that is MJ m^{-2}.

It might seem surprising that the solar energy received at polar latitudes can almost match, at the maximum, the amount received in the tropics (figure 3.3). The reason is the day length at high latitudes: on June 22 at the Arctic Circle—latitude 66°33' N—the sun never sets, giving a 24-hour day; at 60° N in mid-June day length is about 18.5 hours. Even on a clear day, the rate of energy input at high latitudes will be low, but such energy input, maintained for many hours, adds up to a large value.

The relationship between the rate of energy input and the amount received is analogous to the flow of water through a pipe into a tank: water flow rate is measured in liters (or gallons, for Americans) per minute; multiplying the rate by the number of minutes gives the total amount of water that flowed into the tank in a specified time. Solar radiation is measured in the same energy (power) units used to measure electricity use, that is, watts. A watt is a joule per second (J s^{-1}), and a joule is a unit of work; we measure the solar energy received on a surface as watts per square meter per second (W m^{-2} s^{-1}), that is, J m^{-2}. So multiplying the rate by time (in seconds) gives joules,

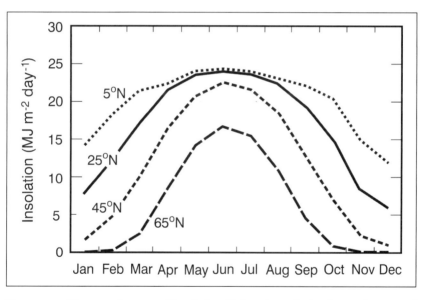

FIGURE 3.3. The annual course of insolation (daily solar radiation income) at several latitudes. The amount of energy received at the ground is strongly dependent on cloud cover and, in the absence of clouds, is affected by aerosols and pollutants. Clouds reduce the amount of energy reaching the ground because they reflect short-wave solar radiation back into space; they also scatter and absorb radiation. Since there is a positive correlation between clouds and precipitation, solar radiation at the ground tends to be inversely related to precipitation. The curves were derived from calculations using a procedure described by Coops, Waring, and Moncrieff (2000).

and since it comes out as a large number over a day, we divide by 10^6 to give megajoules (MJ). Solar radiation input rates in the tropics are obviously higher than in high latitudes, although the highest rates received on earth are in the middle latitudes, in areas where cloud cover and air humidity are low. Input rates are highest in deserts, although they are of very limited interest in relation to forests.

Although the sun's energy drives plant life on earth and is responsible for keeping the earth warm, much of the energy escapes as long-wave radiation. If it did not, the earth would get progressively hotter, and life could not be sustained. This mechanism—of energy loss by long-wave radiation—is well understood and is described by an equation known as Stefan's Law, which says that every body with a temperature greater than absolute zero (−273°C) loses heat by long-wave radiation at a rate that depends on its temperature raised to the 4th power. On clear nights there is a high rate of energy loss into space, and the nights tend to be cold, but when there's cloud cover the long-wave radiation is trapped by the clouds and the nights remain warmer. It's important to understand this process because the rate of heat loss from the earth is being changed by the steady and rapid increase in the concentrations of CO_2 in the earth's atmosphere that has been going on since about the time of the Industrial Revolution in the mid-nineteenth century (see chapter 4).

The visible component of solar radiation—light—sets limits on how fast a forest in any particular place can grow because it provides the energy that drives photosynthesis. Clearly, if leaves are the organs that intercept radiant energy, and the amount of energy intercepted determines how much CO_2 is fixed by photosynthesis, then the amount of foliage, and the way the leaves are displayed by trees, must also be important. Coniferous forests, which generally grow at high latitudes, display their foliage very differently from temperate or tropical deciduous trees. Conifers are also (most of them, anyway: the larches are exceptions) evergreen; that is, they do not shed their leaves each fall and produce new ones in the spring.

Forest ecologists use a measurement called *leaf area index*

(LAI) to describe forest canopies in terms that allow comparisons of their ability to intercept radiation. LAI is the surface area of leaves per unit ground area. It's relatively straightforward to calculate for broad-leaved forests, where the leaves are flat, but coniferous forests, with their needle foliage, are somewhat more complicated: needles are not flat, so projected area may not be a useful measure of total surface area. For needles we use one-half of the total surface area. Determining LAI often involves laborious sampling and measurements in laboratories, but there have been hundreds of studies on it, so there are a great many data sets, for different forest types, available in the scientific literature. There are also instruments available, based on light interception, that allow reasonably accurate measurements of the LAI of forest stands without destructive sampling.

It's possible to establish simple relationships between the amount of light absorbed by forest canopies, calculated from solar radiation and LAI data, and the biomass production rates of forests, and use these to estimate forest growth rates. The relationships must, however, take account of the water and nutritional status of the forest. The relationship shown in figure 3.4 was obtained by Linder (1985), using data from a *Eucalyptus globulus* fertilizer trial at three ages. The proportion of incident radiation absorbed depends on LAI, which was higher for the better-fertilized plots, so growth in those was faster, and it increased with age. Relationships like this can't be expected to continue in a straight line indefinitely, although it's surprising how general it is in that form.

Weather Factors: Air Humidity

Air humidity is important for forests because it is a major driver of the rate at which they lose water. The humidity of the air mass acting on a forest canopy generally depends on meteorological conditions and evaporation from large areas like the ocean, large lakes, or wetlands. Local conditions exert limited effects except that, deep in thick forest stands, air humidity is likely to be higher than

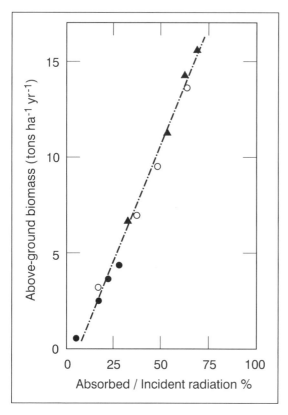

FIGURE 3.4. A linear relationship between aboveground biomass production and the proportion of incident short-wave radiation absorbed by foliage in fertilized plots of *Eucalyptus globulus* at 2 years, (filled circles), 4 years (open circles), and 9.5 years (filled triangles) of age. Redrawn from Linder (1985).

above the canopy. The measure of air humidity that is important in relation to evaporation—and therefore transpiration—rates is the *humidity deficit*. This is conventionally defined as *vapor pressure deficit* (VPD). The standard measure of air humidity that everyone knows is *relative humidity* (RH). RH is the ratio of the amount of water vapor in a given volume of air relative to the amount that the air could hold if it was saturated at the same temperature. Air saturated with water vapor at a particular temperature has RH = 100%. However, RH is not important in relation to evaporation and water use by plants.

The measure of air humidity that is important in relation to rates of evaporation, and therefore transpiration, is vapor pressure deficit (VPD). VPD is the difference in vapor pressure between the saturated vapor pressure of the air at a given temperature, and the actual vapor pressure at that temperature: it's a measure of the drying power of the air. Vapor pressure, as the name implies, is measured in pressure units (pascals [Pa] or kilopascals [kPa] or megapascals [MPa]). For example: the saturated vapor pressure of air at 15°C ≈ 1.7 kPa; if the RH is 50 percent, the VPD is 0.85 kPa. The saturated vapor pressure of air at 25°C ≈ 3.2 kPa; if the RH = 50%, the VPD = 1.58 kPa. (Standard atmospheric pressure is 1,013 Pa, 101.3 kPa, 0.103 MPa.) These values are readily obtained from standard tables.

Air may not be saturated at its present temperature, but if cooled enough it will reach the *dew point* temperature and condensation will occur. For example, the saturated vapor pressure of air at 20°C is 2.34 kPa; if RH = 75 percent, the actual vapor pressure is (0.75 x 2.34) = 1.75 kPa. As 1.75 kPa is the saturated vapor pressure of air at about 15.5°C, that's the dew point of the air: if it is cooled from 20°C to 15.5°C, water will start to condense out of it. The further it's cooled, the more water is condensed. This is the reason for the frost on a cold beer glass and for dew on the grass in the morning, and for fog and mist. Water condenses from warm damp air in contact with the cold glass. Dew is condensed from the air when the surface (grass) cools at night because of heat loss by long-wave radiation. Fog and mist occur when the air mass cools by some other mechanism; for example, mist may sit in valleys on still mornings because cool air has flowed downhill.

Air humidity measurements are usually made using wet and dry bulb thermometers. There are tables and equations available so that the VPD and dew point we have just discussed can be calculated for any temperature and air water content.

The crucial point in relation to calculating water-use rates by forests, or any other plant community, is the assumption, which is usually made, that leaf temperatures are the same as air temperatures. The vapor pressure inside stomata is then saturated vapor pressure at that temperature, and transpiration is driven by the vapor pressure gradient between the leaves and the air. Assuming that leaf temperature = air temperature = 15°C or 25°C, then saturated VP in the stomatal cavities would be 1.75 kPa or 3.2 kPa, respectively. If RH, in both cases, was 50%, the VP of the air would be 0.85 kPa at 15°C and 1.58 kPa at 25°C. These would also be the values of the VPD, so the evaporative demand would be double at the higher temperature, although the RH is the same.

Weather Factors: Precipitation and Hydrology

Precipitation usually means rainfall, but snow and hail have to be included in assessments of what is called the soil water balance. In what follows we mainly consider rain; the principles are the same for snow or hail, but the details are different. For example, a forest canopy can hold about twice as much water in the form of snow as it can in liquid form. Hydrology is not in itself a weather factor; it's actually the science of water movement across the earth, but since it's very much about precipitation and what happens to it, we have chosen to treat precipitation and hydrology together rather than separately.

Rainfall is measured in depth units: the number of millimeters (or inches, in America) of rain that fall in an "event" (the jargon conventionally used) indicates the depth to which that rain would cover the floor of a flat container with vertical sides. We measure rainfall with rain gauges that are not flat—there would be too many problems with splash and evaporation (not to mention animals that might fancy the gauge as a watering trough)—but convert the amount measured into depth equivalents. There are some interesting and handy conversions associated with this sort of thing: 1 liter of water covers 1 square meter to a depth of 1 mm. Therefore, if we know how much rain (in mm) has fallen in an event over a known area, it's very simple to work out how much water could be collected, or could run off the surface or enter the soil. For example, a 10 mm fall over an area of 1 square kilometer (1,000 x 1,000 m) is 10 million liters of water.

We have not provided tables or graphs of precipitation amounts in various regions: obviously the numbers vary enormously across the world, from high-rainfall, wet tropics to dry deserts. The average amount of precipitation that can be expected annually at any location depends on global circulation, topography, proximity to oceans or very large lakes. Forests do not grow in dry areas of the world; as a rough rule of thumb, they do not occur in areas where the average annual precipitation is less than about 700 mm per year, although that depends on the seasonal distribution of rainfall in relation to evaporation, as well as on the capacity of the soils to store water.

Average annual rainfall, although a useful and commonly quoted statistic, is really only an indicative number. The seasonal distribution of precipitation through the year is also important: if there are definite wet and dry periods, there may be a significant probability of drought. People are often surprised to find that much of the Amazon basin, which we're inclined to think is always wet, has a dry season (refer back to figure 2.4). An interesting climate, in this respect, is the Pacific Northwest area of the United States, which experiences cool, damp winters but consistently hot, dry summers. That area is also interesting because there is a spectacular gradient of precipitation from the coast (high rainfall), to the Cascades (high snowfall) to the eastern rain shadow where, only a couple of hundred kilometers from the coast, the country is semiarid. The coastal ranges of the Pacific Northwest contain some of the largest trees in the world, and the transect includes a tenfold range in LAI and forest productivity.

Precipitation affects the growth of plant communities primarily through its effects on the amount of water in the soil, so it's these effects, and their implications for tree growth, that are the main concerns of forest ecologists in relation to precipitation. Most trees grow best when the soil round their roots is wet but not saturated. (Saturated soil is likely to be deficient in oxygen, which is needed for roots to function effectively.) Hydrologists are also concerned with the amount of water stored in soils, but their main concern is how much runs off or drains through soils from a particular area. *Drainage* and *runoff* are, of course, major factors influencing stream and river flow.

Hydrologists and forest ecologists calculate the water balance of areas of any size—small or large *catchments*,[3] forest blocks or regions—using an equation that is conceptually quite simple. It's called the *hydrologic equation*, and it can be written thus:

Change in soil–water content across a time interval = Precipitation – intercepted water (evaporated) – runoff – drainage out of the soil – water transpired by plants. (equation 1)

The equation is called a conservation equation and it describes what happens over some time interval to all the water that reaches the earth (which includes the canopies of trees) as precipitation. The equation is usually written in symbolic mathematical form, so it can be manipulated, but it's perfectly accurate as written here. The time interval can be in any convenient units—hours, days, months—but for purposes of discussion we will assume that we're talking about an interval of a day. Our problem is getting the right values for all of the terms, but what matters most to trees is the actual soil water content at any time—the first term on the left-hand side. The *hydrologic cycle*, as it's called, is illustrated in the diagram in figure 3.5.

We'll discuss the terms in the hydrologic equation in the order they are written here, but note that it is difficult to completely separate the various terms, since the processes they describe may be operating simultaneously. The implications of forest cover in catchments, and its effects on water quality and the water yield (the amount of water that will come from a catchment relative to the amount of precipitation that falls on it) are discussed in chapter 5.

Soil Water

What do we mean by soil water? Soil is a porous medium—a layer of soil of any specified thickness can hold a finite amount of water. In both ecology and hydrology we deal with specified depths of soil; in forest ecology we are usually concerned with the depth exploited by the roots of the trees—the root zone (defined earlier).

There's an upper limit to the amount of water that a soil can hold;

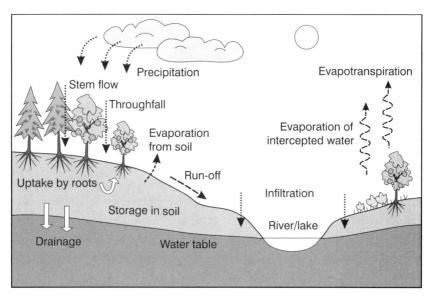

FIGURE 3.5. Diagram showing the flows of water and water vapor involved in the hydrological balance of a forest stand or catchment. The amount of water stored in soil exploited by tree roots (the root zone) at any time depends on the balance between inputs by precipitation and losses from runoff, drainage out of the root zone, evaporation of intercepted precipitation from wet soil and foliage, and transpiration. Water stored in the root zone is removed primarily by transpiration. If there is no precipitation during the time interval of interest, some water may move into the drier volumes from wetter areas, where the source may include a water table.

if it's saturated with water, any addition will either run off the surface or drain out of the root zone until a point is reached where gravity can no longer overcome the adhesive molecular forces between water molecules and the solid particles of the soil. This upper limit of water that a soil can hold against gravity drainage is called *field capacity* (FC). It varies considerably between coarse-grained sandy soils and fine-textured soils.

There is a low moisture-content limit at or below which plants cannot extract water from the soil. This is called the *wilting point* (WP), referring to the fact that plants with broad, soft leaves (like tomatoes, which were the original plants used in research on these matters) wilt in soil at that water content. The WP defines the water content at which all plants stop extracting water from the soil. It doesn't mean the soil is completely dry; that state is almost never reached. For ecological, agricultural, and hydrological purposes, the water-holding capacity of soils is the amount of water that can be held in the root zone between FC and WP.

To give a few representative figures for the water-holding capacities of soils: fine-textured soils, with a large amount of clay in them, can hold much more water than sandy soils. They have high *porosity*, which is the ratio of the volume of pore space per unit volume of soil: about 0.5 in a fine-textured soil with high clay content. So at saturation, about 50 percent of the volume (denoted 0.5 v/v) of a clay soil will be filled with water. At FC it will be about 45 percent. Plants may not be able to extract water from the clay soil when the water content falls below 30 percent—so the soil is still quite wet at the WP, because the soil holds the water very tightly. A sandy soil may have porosity of 0.3–0.4. Over 40 percent of the volume of a sandy soil may be filled with water at saturation, but only about 10–15 percent at FC. This very large difference reflects the fact that sandy soils are free draining: they don't hold water very tightly. For this reason plants can get water out of the sandy soil down to much lower water contents; the WP may be at about 5–8 percent by volume.

Those who like things quantitative may wish to explore the implications of those numbers. So let's assume a forest where the trees have most of their roots in the top one meter of soil. The arithmetic is straightforward if we take a land area of one square meter, so we are considering a block of soil with a volume of one cubic meter ($1 m^3$). At FC, $1 m^3$ of a clay soil holds about $0.45 m^3$ of water; since $1 m^3 = 1,000$ liters, that is 450 liters of water. At WP the soil may still contain 30 percent (by volume) water, that is, 300 liters. The difference—150 liters water per square meter (liters m^{-2}) of land area—is available for plants to use. By contrast the sandy soil holds, let's say, 12 percent v/v water, that is, about 120 liters in the root zone at FC. At the WP, say, 5 percent v/v, the amount of water in the root zone is 50 liters, so the amount of water available for the trees to use is 70 liters—about half the amount available from the heavier soil. These numbers indicate the amount of precipitation that soils can accept, which will, of course, depend on how wet they are to start with, as well as their porosity and the depth to which roots penetrate.

So it's immediately obvious that a heavy clay soil, with a water content near the WP (0.3 v/v = $300 \, l \, m^{-3}$) in the top one meter, will be able to accept about 200 liters of water per square meter of land area ($200 \, l \, m^{-2}$) before saturation at $500 \, l \, m^{-3}$, that is, since 1 mm rain is 1 liter m^{-2} this soil can absorb 200 mm of rain (for those who think in inches, that's almost 8 inches). So if the precipitation is 85 percent efficient, that is, losses from interception by the canopy and understory are 15 percent, then to get 200 mm into the soil will probably require about 235 mm of rain. Similarly, a dry sandy soil at WP could accept 40–50 = $350 \, l \, m^{-2}$ before saturation, that is, 350 mm (about 13.8 inches) rain. But if FC is only, say, 0.15 v/v, that is, $1.50 \, l \, m^{-3}$, then 200 liters of that water would drain quite rapidly out of the root zone. It may take a couple of days, depending on things like subsoil conditions, but drainage won't stop until the soil reaches FC.

Precipitation, Interception, Runoff, and Drainage

We have, inevitably, in the previous paragraphs, touched on the matter of drainage out of the root zone. We deal with drainage and runoff and the implications of those processes in a little more detail later in this section.

When rain or snow falls on a forest, some of it is intercepted by foliage and bark and never reaches the ground, so the effective precipitation is the amount that reaches the ground. The water held on foliage is evaporated back into the atmosphere. While foliage is wet, transpiration is virtually zero, and small amounts of the surface water may enter the leaves if stomata are open. The proportion of any rain event that is intercepted depends on the intensity of the rain, on the density of the foliage and the type of event. Obviously a dense canopy will hold more water than a sparse one. If the rain is pouring down, once the foliage is wet, virtually all the rest reaches the ground, some of it directly, some running down stems. If the rain is low intensity, slightly more will be held by the foliage because the heavy drops of high-intensity rain shake twigs and leaves and heavy rainfall is more likely to be accompanied by wind. Intercepted water will be lost by evaporation, which does take place during rain, so that more water will then be intercepted. In the case of intermittent light showers, which wet the canopy and dry off before the next one, a very high proportion of the rain above the canopy may never reach ground level. Finally, not everything that gets through forest canopies reaches the soil; there is usually a layer of litter and dead material, and small plants known as understory, which also intercept water and from which it may evaporate. A great deal of research has been done on all this and, given rainfall measurements and information about forest canopies, we can make reasonably accurate estimates of the amount of rain that will be lost by interception.

Snow is a bit more complicated than rain because much more reaches the ground in open areas than under a canopy. But evaporation (actually *sublimation*—the direct transformation of snow to vapor) is much higher in the open than under a canopy so that the amount of water entering the soil is similar, except for the timing. One of the important differences between the interception of rain and snow is the (so-called) residence time of snow on the canopy. Snow may remain on the canopy for long periods (days, weeks) depending on factors such as wind speed. Also, in areas with cold winters, snow accumulates on the ground during winter so that, when it melts in the spring, a large amount of water is released.

Melting snow is different from heavy rain in that the process is slower, but it replenishes soil moisture reserves and is also likely to result in large amounts of runoff. More snow is likely to accumulate, and remain longer, under forests than on ground covered by low vegetation such as rangeland, or fallow agricultural fields, from which much of it will sublimate or be blown away during the winter. The hydrology of large areas of the world is dominated, or at least heavily influenced, by snow and snowmelt. This includes much of the western United States, where the snow on the mountain ranges serves as a major reservoir for much of the year, and snowmelt is a vital source of water.

Runoff and drainage are both loss terms as far as plants are concerned, although of vital interest to hydrologists. Forests use more water than agricultural crops and grassland, which may seem to provide a reason, in this increasingly crowded world where water shortages are developing in many places, for removing forests. And indeed, the total runoff from cleared land may be more than from forested land, but that runoff will come in surges that may cause flooding, while during rainless periods, river flows will be low. High runoff rates may also cause damaging soil erosion and, where dams have been built for water storage, this will contribute to those dams silting up. Forests act as sponges that hold water and release it slowly. Water penetration into soil under forests is generally much faster than when the soil is covered by grass or crops or, even worse, is paved (no penetration) or heavily compacted. Litter layers on the forest floor slow the rate at which water runs off across the surface and allow water to seep into soils, even when they are saturated, at rates commensurate with drainage through the root zone. It is that slow drainage out of the root zone and into the subsoil and laterally underground that feeds springs and streams and keeps them flowing through dry periods. It also contributes to the groundwater stored deep in soils.

Transpiration

The last term in equation 1 is water transpired by plants. This is a major and very important component of water use by any plant community, including forests. Strictly speaking, the term *water use*—or evapotranspiration—includes the evaporation of intercepted water and evaporation from wet soil, in fact all the processes that return water to the atmosphere as water vapor. However, water use is frequently taken to mean *transpiration* alone, and we will use it in that sense.

Transpiration is the process of evaporation of water from the substomatal cavities in foliage. Tree water use is the sum of the transpiration from all the leaves on a tree, and the water use of a forest is the sum of the transpiration, over any specified time interval, of all the trees and other vegetation on a specified area of land. For forests with the same LAI, the rates of transpiration, determined by the *evaporative demand* of the atmosphere, are obviously likely to be higher in hot, dry areas than in cool, damp areas.

The evaporative demand determines the rate at which water will be evaporated from forest canopies where the roots have access to adequate soil water. It depends on the VPD of the air, radiant energy load on the foliage, and wind speed, which determine air flow rates across leaves. The process is analogous to clothes drying on a line: they will dry more quickly on a hot day with low humidity and on windy days. We will not be elaborating on energy interception and wind speed, except to note that the effects of both are modified by canopy density, measured by LAI. If VPD is high, tending to drive transpiration faster than water can be moved from soil to roots and through the conducting system, the stomata of most tree species will close, wholly or partially, reducing the loss rates and preventing the irreversible damage that could result if leaves became desiccated. This is why stomatal apertures change with air humidity and why they generally close proportionally to VPD increases during the day; as we noted earlier, stomata "fine-tune" the responses of trees to water loss rates.

Transpiration takes place mainly during the day, and it will be faster from forest canopies with large amounts of foliage, at least up to a limit. Beyond a certain foliage density, corresponding to an LAI of ~5, additional leaves will not make much difference because the increased density of leaves in the canopy results in increased mutual shading—and reduced energy load on shaded leaves. Lower light levels also prevent stomata in shaded leaves from opening widely. This effect of high LAI also explains why dense forests, with LAI >6, can be heavily thinned without measurably reducing wood production: maximum stand photosynthesis rates are reached at about that value.

To give some "ball park" figures for maximum forest water-use rates, the highest rates measured in research projects are of the order of 5 mm per day, in eucalyptus plantations in Australia. (These figures are exactly analogous to rainfall; they represent a depth of water.) More usual summer rates in temperate regions are around 2 to 3 mm per day while, because of the very high air humidity in the wet tropics, forests there only transpire about 1 to 2 mm per day. Tying these figures into soil water-holding characteristics, we gave a figure of 150 mm of available water for a clay soil and about 70 mm for a sandy soil (assuming a root zone 1 m deep). Starting at field capacity, trees on the clay soil, transpiring at 3 mm per day could therefore keep going for fifty days without rain, while trees on the sandy soil would only have water for about half that time at maximum rates. Of course transpiration rates are not constant, since they depend on day-to-day weather and constraints imposed as the available water supply in the rooting zone is reduced.

Implications of Forest Clearance for Hydrology and Climate

When forests are removed, surface flow rates tend to be greatly increased, so that heavy precipitation, or snowmelt in spring, results in the rapid runoff of large volumes of water, increasing the likelihood of flooding. This can be seen in various areas of the world where forest cover has been destroyed. An important example is Nepal where the Ganges River rises in the slopes of the Himalayas. Many of those slopes have been denuded of their forest cover over the years as the Nepalese population has increased, bringing increasing pressure on the land, which is cleared by woodcutters and for grazing. The result is a dangerous change in the flood patterns of the Ganges, affecting millions of people who live along the river and in its estuary, Bangladesh. The dangers are exacerbated by the melting of the glaciers in the Himalayas (and virtually all over the world, including the famous Glacier Park in the Rocky Mountains), possibly associated with global warming.

Clearing forests for agriculture may change hydrological patterns in more ways than runoff and river-flow patterns. In the wheat belt of Western Australia the replacement of native forests by agricultural crops has resulted in the development of a major hydrological problem. The removal of the perennial, deep-rooted trees and their replacement by crops that use less water, from a shallower root zone, and are only in the ground for part of the year, has allowed water tables to rise, bringing with them salt stored deep in the soil. The result is soil salinization, so that the land, in many places, has become unusable. We could cite many other examples.

There is one more consequence of large-scale forest removal that should be borne in mind, and that is the change to the energy balance of denuded areas. We outlined earlier the relationship between the temperature of a body and its loss of heat in the form of long-wave radiation, and noted the implications of the fact that the interception of long-wave radiation by CO_2 is the major mechanism driving global warming. Much of the short-wave radiant energy that strikes forests is absorbed and converted into water vapor, but if the forest in a tropical area is removed, then, depending on the state of the new surface, it is very likely that the surface will dry out and become significantly warmer than the forest would have been. This means more long-wave radiation is emitted into the atmosphere, some of which will be absorbed by greenhouse gases. Also, the temperature of air depends, to a large extent, on the temperature of the underlying surface—heat is transferred to the air from the surface. The warmed air may be moved around laterally by wind, but it still contributes to the warming of the atmosphere. On a large scale this effect may be considerable: it has been calculated that, if the whole Amazon basin was cleared and converted to crops and grassland, it would make a significant contribution to global warming. (It would also radically change the hydrology of the Amazon basin.) This is just another reason why we must hope that the alarming clearing of the Amazon forests can be halted.

We should note here that, while we have outlined the (generally adverse) implications of clearing forests, there is another side to the story, concerning the effects of plantations on the hydrology of catchments. In South Africa the establishment of eucalypt plantations has been restricted in some areas, particularly on the upper slopes of catchments, because they have altered the hydrological balance, reducing stream flow and hence the water supplies to villages in the lower parts of the catchments. Eucalypt stands and plantations are also regarded unfavorably in some parts of India, for similar reasons.

Summary

This chapter provides basic background information about the physiology of trees and how weather and climate act on them, and interact with their physiology, to determine how trees grow, and what characteristics enable different tree species to grow in different climatic regions.

Roots, stems, and leaves are common to all trees. Stomata, small pores in leaf surfaces, are at the interface between the environment and the physiological processes that underlie growth: carbon dioxide is absorbed through stomata and used for photosynthesis, the process by which carbohydrates are produced, using the energy in the visible wavebands of solar radiation. Water vapor is inevitably lost, by transpiration, through the stomata, and has to be restored by uptake from the soil through the roots.

The weather factors affecting growth are temperature, which determines the rate of physiological process; short-wave solar radiation, which drives photosynthesis; and air humidity, which affects stomatal apertures and transpiration rates. Temperatures vary with latitude, season, and topography. Solar radiation, which also varies with latitude and season, is absorbed by foliage and is the major factor determining tree growth. The energy balance of the earth depends on the balance between incoming solar radiation and the energy lost by long-wave (heat) radiation. Air humidity reflects the amount of water vapor held in the air, which is strongly dependent on temperature.

Plants need water. The growth of plant communities is highly dependent on the amount of water in the soil, which, at any time, depends on the (hydrologic) balance between precipitation and losses by runoff, evaporation, drainage, and transpiration. This balance is influenced by the capacity of soils to store water.

Forest clearance has generally adverse effects on hydrology and climate.

Chapter 4

Causes and Consequences of Rapid Climate Change

The earth's climate has never been stable; geologic records show that, over millennia, the changes have been massive, ranging from ice ages to warm periods. But, within the last hundred years, as a result of human activities, air temperatures around the globe have been increasing at an unprecedented rate. This has serious implications and consequences, not just for forests but for all natural ecosystems and for human societies. Biological organisms and ecosystems have been able to change and adapt to the slow changes in earth's climate that have occurred in past geologic ages, but they cannot change fast enough to adapt to the pace of the changes happening now.

Rising global temperatures are triggering changes in the frequency and intensity of all sorts of other events, such as droughts, floods, hurricanes, the melting of glaciers and of the summer ice on the Arctic Ocean. It's probably too late, now, to reverse the global warming process, although we must try to ameliorate it by reducing the activities that are causing climates to change. We must also, based on an understanding of the causes and effects, work to adapt the way our societies use the resources of the world and interact with natural systems. Forests are key factors in all this: they are affected by climate change and also help ameliorate it. Forests are heavily impacted by human demand for resources and therefore, as human populations grow, they come under more and more pressure. The whole question of climate change is therefore highly pertinent to this book.

Causes of Global Warming

John Tyndall established, in Britain in the nineteenth century, that methane and carbon dioxide control the earth's temperature by absorbing *long-wave* or *infrared (heat) radiation*. The atmosphere is transparent to short-wave solar radiation, which warms the earth but, as we noted in chapter 1, much of the energy absorbed by the earth escapes as long-wave radiation. If it did not, the earth would get progressively hotter and life could not be sustained. But a proportion of the long-wave radiation is absorbed by carbon dioxide (CO_2) and other "greenhouse gases" in the air, such as methane, nitrous oxide, and water vapor, and some of it is reradiated back to earth. This finely balanced "blanket," or *greenhouse effect*, is extremely important for maintaining the stable temperatures suitable for life on earth. But if the concentrations of absorbing gases increase, the amount of long-wave radiation absorbed in the atmosphere and reradiated back to earth will increase, and temperatures at the ground and in the lower layers of the atmosphere will rise. This is called *radiative forcing*, because the mechanism of temperature change depends on changes in the radiation balance. In 1896 the Swedish chemist Svante Arrhenius provided an equation describing the radiative forcing effect of CO_2 that is still used today.[1] He was the first person to predict that emissions of CO_2 from the burning of fossil fuels and other combustion processes could lead to global warming. A change in the earth's energy balance of 0.3 percent, equivalent to a change in radiative forcing of 0.5 W m^{-2}, is enough to flip us into or out of an ice age.

An American scientist—Charles Keeling—started a series of measurements of atmospheric concentrations of CO_2 on the

Hawaiian island of Mauna Loa in 1958. The observations were made at an altitude of more than 3,000 m, above the inversion layer so they provided a good sample of the air over the northern Pacific Ocean. By 1960 Keeling had established that there were strong seasonal variations in atmospheric concentrations of CO_2, with peaks in the Northern Hemisphere winter. The measurements that Keeling began have been continued to the present day and provide the longest record of atmospheric CO_2 in the world. They show that concentrations have risen from around 315 parts per million (ppm, by volume) in 1958 to around 400 ppm now. (Ice core data show that CO_2 levels have not exceeded 300 ppm for the last 650,000 years.) As Keeling's results became known, and similar measurements from various sites around the world confirmed the steadily rising concentrations of CO_2 in the atmosphere, scientists became concerned that these would cause changes to the earth's climate. We outline, in the next section, the reasons for this concern. Most scientists now accept that these changes are the primary cause of the current rapid increase in global temperatures. The changes in atmospheric CO_2 concentrations from the beginning of the twentieth century to the present day are shown in figure 4.1.

Carbon dioxide, as we said earlier, is not the only gas that affects the earth's radiation balance. A molecule of methane is more than 20 times as effective at trapping radiation as a molecule of CO_2, and nitrous oxide (NO) is 300 times more effective as a heat-trapping gas than CO_2. Humans are responsible for considerable increases in methane emissions: cattle produce a great deal of it, and there are enormous and increasing numbers of cattle in the world.

Methane is also produced by the rotting waste in landfill and emissions from wetlands and melting permafrost in the Arctic. Boreal peatlands contain a large proportion of the world's soil carbon because plant material, produced over hundreds of years and submerged in the high soil water tables, is incompletely decomposed, resulting in accumulation of carbon as peat. Slow microbial decomposition and anaerobic respiration result in the emission of methane. This is likely to be accelerated by rising temperatures.

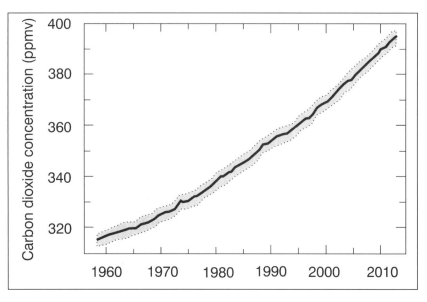

FIGURE 4.1. Changes in CO_2 concentrations in the earth's atmosphere in the modern era, based on the measurements made on Mauna Loa in Hawaii. Redrawn from data of Tans and Keeling (2013).

Other gases, such as chlorofluorocarbons (which have been banned in much of the world because they also degrade the ozone layer), have heat-trapping potential thousands of times greater than CO_2. But because their concentrations are much lower than CO_2, none of these gases contribute as much to global warming as CO_2. Carbon dioxide also has the property that it is conserved in the atmosphere for hundreds of years; it doesn't break down, so once it's there, it stays there, and its effects continue.

Fossil Fuel Burning and Land-Use Change

So where is all the additional CO_2 coming from? Human activity causes the emission of greenhouse gases into the earth's atmosphere in a variety of ways. Most of the emissions come from burning fossil fuels such as coal, oil, and natural gas to provide the energy that powers the modern world. There are thousands of coal-fired power plants round the world—about 620 in China alone. The shift to the use of natural gas, which emits about half the amount of CO_2 per unit energy produced, is helping to slow emission rates, particularly in the United States, but coal remains the primary fuel used in most Chinese and Indian power stations and across the (so-called) developing world. The other major sources of emissions are transport—trucks burning diesel fuel are the most important contributors of both CO_2 and particulate pollution—and cement manufacture and land-use change. The process of cement production, an enormous industry, particularly in China, India, and the United States, produces massive amounts of CO_2 because it's based on heating calcium carbonate, which produces lime and CO_2. Generating the energy used in the production process also involves the emission of large quantities of the gas.

The other major source of CO_2 emissions is land-use change. Any land clearance, with associated vegetation burning and cultivation, releases into the atmosphere as CO_2 the carbon stored in the plants that are destroyed in the process. (Data on the amounts of carbon stored in the world's forests, and the annual fluxes of carbon into and out of them, are given in the next chapter, in table 5.1.) Agricultural activities in themselves release more CO_2 as carbon stored in the soil is oxidized when the soils are cultivated. Land-use change is driven by increasing human populations and the constantly increasing pressures for higher standards of living, which inevitably involve more and more consumption. Much of the increase in human populations is in the developing world: feeding the people is becoming increasingly difficult and a United Nations report suggests that almost 50 percent of deforestation is a direct result of subsistence farming. Throughout the tropics, driven by rising human populations, land is being cleared for agriculture. People in South America, Africa—particularly west and central Africa—and Southeast Asia use primitive *slash and burn agriculture* techniques to push into the forests and expand the areas available to them for food production. Wherever forests are penetrated by roads, people will follow. In Brazil, in particular, deforestation and land-use change can be seen from the satellite photographs that show the "herring bone" patterns along roads driven into the forests for (frequently illegal) mining development or logging operations (there is further discussion on this in chapter 5).

Much of the organic carbon stored in forest soils decomposes rapidly, emitting CO_2, when the forests are disturbed. A large proportion of the world's soil carbon is, as we pointed out earlier, contained in the peatlands of the boreal regions and, whether the forests there are physically disturbed or not, methane emissions from those regions are being accelerated by rising global temperatures.

The other major source of carbon emissions from land-use change is deforestation, either to use the land for commercial cropping or as a result of uncontrolled logging. About 13 million hectares (130,000 km²) of land are deforested each year (Canadell and Raupach 2008), mostly in tropical regions, releasing about 1.5 billion tons of CO_2 (see figure 4.2). Clearance for commercial cropping has taken place on a massive scale in the Amazon basin, largely for the cultivation of soybeans that are exported primarily to the United States as stockfeed. (A confronting article on this subject can be found in *National Geographic Magazine*).[2] About 20 percent of the Amazon rainforest has been destroyed in the last forty years. Rapid forest destruction is taking place across Southeast Asia and the Pacific region. The worst of the destruction is in Indonesia, where there has been massive forest clearing and biomass burning in recent years, largely for oil palm establishment (see chapter 5 for more details). The fires result in smoke haze across Indonesia itself, and Malaysia and Singapore, with significant effects on tourism and public health. The main effect, in relation to climate change, is the input of CO_2 into the atmosphere. The local effects on radiation balance and hydrology are likely to be much smaller than in the Amazon, except immediately after clearing and burning, because oil palm plantations, being populations of trees, respond to incident radiation like natural forests.

Deforestation not only results in the release of large amounts of CO_2, but it also changes the energy balance of deforested areas

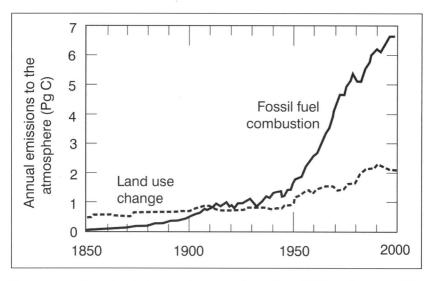

FIGURE 4.2. Annual emissions to the atmosphere, from 1850 to the year 2000, as a result of fossil fuel combustion and land-use change. The large increase in the contribution from land-use change and the associated burning and decomposition of biomass since 1950 indicates that this has been a significant contributor to total global emissions since that time. Forest clearance is a major factor in this.

because it results in changes in the *albedo*—the proportion of short-wave energy from the sun that is reflected by the underlying surface. The albedo of full-canopied, forested land is about 0.1–0.12, that is, it reflects about 10 percent to 12 percent of the short-wave energy incident on it. If the forest is removed and the land converted to a crop, such as soybeans, the albedo rises to about 15 percent to 18 percent while the crop is in the ground, and may go higher when it dies off and is harvested. The albedo of bare soil depends on its color and wetness; it may vary from as low as 10 percent for a wet, black soil to more than 30 percent for light dry soils. When a crop is removed, the soil surface tends to dry out and reflect an increased proportion of incident radiation. Up to now, changes in the regional energy balances caused by changes in land use are small enough to be difficult to detect on a global scale, where they are swamped by the influence of greenhouse gases, but they will become increasingly significant if forest destruction continues.

The development of agriculture on land cleared of forests for that purpose not only releases large amounts of CO_2, but it also changes the hydrology of the affected areas. The rate at which water is transpired from a low crop is always lower than the rate from a healthy green forest. Transpiration converts heat and radiant energy into latent heat, that is, water vapor, so the rate of heat transfer to the atmosphere is low from an actively transpiring vegetated surface such as a forest. The lower rate of water loss from cropped surfaces or cleared areas relative to forests means higher rates of heat transfer into the air. When the crop dries out, rates of evaporation from the surface may fall to very low levels, except immediately after rain, and the rate of heat transfer to the atmosphere goes even higher. As greater areas of forest are replaced with cultivated land, there is the danger of significant *feedback effects*: higher local temperatures cause increased rates of water use by the trees in the area. These, in turn, make the trees more vulnerable to droughts caused by periods of low rainfall, so the probability of tree death increases. At some point the system will reach a regional *tipping point*: if enough rainforest is cleared there will be widespread tree death, changes in the type of vegetation, and yet further exacerbation of the climate change effects.

How Fast Is the Earth Warming?

The question of how fast the earth is warming as a result of the increases in greenhouse gas concentrations in the atmosphere is being addressed in two ways. The first, and obvious, approach is through measurement. Air temperatures have been measured in England since the seventeenth century, and at a number of other places around the globe since the nineteenth century. This is known as the instrumental record. It provides an immensely valuable resource, although it must be treated with care, because of differences in instrument quality and exposure and differences in recording procedures. The measurement of air temperature can be a surprisingly inexact business, not because we can't measure temperature accurately but because air temperature varies quite markedly from place to place—for example, on opposite sides of a hill or with distance from a large city, or in a forest clearing, depending on the size of the clearing. But the instrumental record is constantly being refined and improved and observing stations in every country now make available a vast database of temperature records. Measurement techniques and procedures are becoming increasingly standardized, making it much easier to analyze and compare data. Ground-based measurements are supplemented by temperatures measured at sea, from permanent buoys and ships, and at various altitudes from aircraft and weather balloons. Atmospheric temperatures can also be inferred from satellite measurements of atmospheric radiance, which varies with the temperature of the gaseous components.

The second approach to determining air temperatures across the globe is to predict them using *global climate models* (GCMs). GCMs are not about predicting the exact weather at any time or place in the future, but rather about projecting the statistics (average behavior and variability) of the future weather (Pittock 2005). They are based on the general circulation models that are at the core of modern weather forecasting. They are mathematically complex and use a great deal of computer power to simulate the general circulation of the earth's atmosphere and the interactions between the atmosphere, land, and water surfaces. GCMs include the effects of sea surface temperatures and land surface conditions in terms of albedo and radiation balance, as well as atmospheric conditions. The inclusion of the causes of accelerated global warming in the global climate models allows exploration of the consequences of this warming in terms of various scenarios: we can see what is likely to happen under different conditions and assumptions.

Climate models serve two important purposes: the first relates to the underlying rationale of most scientific modelling—the need to understand the way a system under study "works" in terms of the processes that determine its behavior and responses to stimulus (which might mean any sort of change). The second is prediction. You can predict the future behavior of a system by extrapolating from historical observations, but extrapolation implies the assumption that the system will continue to behave in the future as it has in the past, and that the inputs—the factors that govern its behavior—will be the same. This, by definition, is clearly not a good assumption when we are dealing with climate change. Prediction using models based on processes allows exploration of the consequences of changing the input conditions. These models still involve assumptions and embody uncertainties, but provided they are soundly based on well-understood processes and tested in terms of the factors that affect those processes, we can use them with confidence.

GCMs have been developed at a number of institutions around the world, and there is a high level of collaboration between those institutions. The models are constantly being compared, tested against one another, refined, and improved. They are tested against historical data and variables such as seasonal fluctuations in temperatures or rainfall. They are also tested against the long-term variations in climate revealed by studies based on ice cores, geological strata, and indirect measures of climate change. An enormous amount of scientific effort is going into all this. It is essential that the credibility of the models should be, as far as possible, beyond question so that they provide the analytical and predictive tools we need to confront the implications of climate change and (hopefully) make policy decisions about how to mitigate and adapt to it.

GCMs are now widely used to estimate the temperature changes likely to result from given concentrations of greenhouse gases. Those changes are not, and will not be, uniform across the globe, but the models can be used to predict how much change is likely in particular regions. They also provide predictions about the amount and distribution of precipitation, and are increasingly used to analyze extreme weather events. Changes in the earth's temperature are usually presented as differences in relation to some standard baseline, called the temperature anomaly. The basic fact is that there has been an increase in average global air temperatures of about 0.8°C between 1910 and the present day. The baseline for the data shown in figure 4.3 was taken to be about 14°C.

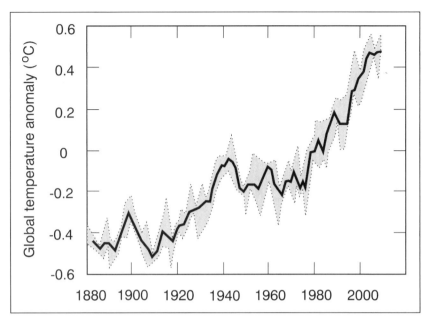

FIGURE 4.3. Global temperature anomaly (i.e., temperature differences from a standard baseline, in this case about 14°C) between the years 1880 and 2000. The data come from the instrumental record and were collated by the Intergovernmental Panel on Climate Change. Redrawn from the website http://en.wikipedia.org/wiki/Instrumental_temperature_record.

Many analyses have been done to try to answer the most important question arising from data such as those shown in figure 4.3: that is, what can we expect in the future? The analysis produced by the United States Environmental Protection Agency shows typical results, presented in figure 4.4. The graph illustrates the probable consequences of three different emission scenarios: low, relative to present emission rates; similar to present rates; and high emission rates. These are usually equated to economic activity. The low rates could come about as a result of reduced fossil fuel and biomass burning—possibly, but not necessarily—correlated to reduced economic activity but also possibly, and quite feasibly, the result of increased energy use efficiency round the world. The middle line is the "business as usual line," taking into account the likely increases in economic activity and the associated fossil fuel

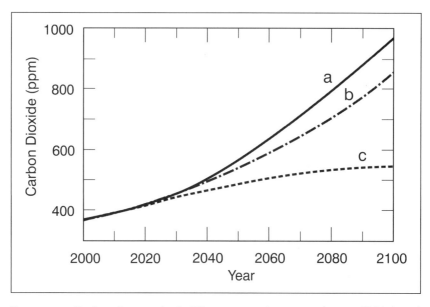

FIGURE 4.4. Projected atmospheric CO_2 concentrations up to the year 2100, based on analyses of global emissions, assuming increasing emission rates (curve a), rates maintained similar to those at present (curve b), and reduced rates (curve c). Even if we can reduce emission rates, atmospheric CO_2 concentrations will continue to increase. Redrawn from the website http://www.epa.gov/climatechange/science/future.html.

consumption that will follow increasing human populations and resource consumption. The high rate assumes increased emissions. We note that, under these scenarios, projected emission rates are likely to lead to increases in global average temperatures of about 3°C, 5°C, and 6°C, respectively, by the end of this century. We also note that recent studies in the boreal zones indicate that permafrost is starting to melt. It is calculated that if the earth warms by another 0.8°C, about eight hundred gigatons of CO_2 trapped in the Arctic bogs and peat will be slowly released. This is about twice the amount currently contained in the atmosphere and could create a positive feedback—more warming releases more methane and CO_2 into the atmosphere so the rate of warming increases.

No one would claim that these projections are certainties. They are based on models that contain assumptions and incorporate assumptions about greenhouse gas emissions. The actual temperature increases may be significantly different from those shown. In the best case, the increase at the end of the century may be smaller than 3°C; in the worst case, it may be double that. Regardless of exactly how big the changes resulting from climate change are likely to be, they are already having major adverse impacts, and these can only get worse. We have to take action to mitigate and adapt to climate change. The actions we take must be based on clearheaded and objective evaluation of the risks associated with global warming and its effects, and the consequences of those effects. They must include the development of policies and patterns of behavior based on objective evaluation of those risks, and on the precautionary principle: we are not sure about exactly how big the changes and their effects will be, but we should take action on the assumption that, if we don't act now things could get very bad.

This book is about forests and forestry, so we are not concerned here with the implications and possible impacts of climate change across all walks of life. Rather, we are concerned with its impacts on forests and the contribution forests can make to reducing the rates of temperature increase, and the impacts of the resulting climate change across the world. We will focus on these matters in the next sections, but before we do so we should consider the question of climate change denialists.

A Word about Denialists

Almost all scientists accept the evidence for, and reality of, anthropogenic (human-induced) climate change. As the evidence for global warming has become more and more difficult to ignore, over the last thirty years, the number of people who assert that it's all nonsense and that we (society as whole) need not do anything about it—certainly not take "heroic" actions that might involve major changes in the way our economies and societies function—has fallen steadily. But there are still denialists, some of whom are very vocal. We need to consider the basis of their dissent.

Denialists argue that the fact that scientists are prepared to acknowledge uncertainties indicates that their knowledge and/or databases are uncertain and therefore the projections can be dismissed. There is no validity in this argument—if it was taken to its logical conclusion and applied to everything in life the conclusion would have to be that we would never, for example, insure ourselves (or our cars or houses) against sickness, accidents, or incidents on the grounds that they may not happen to us. A more sensible approach is to assess the risk of illness or accident (the job is done by actuaries) and buy insurance commensurate with the risk we are prepared to tolerate and the consequences of accidents. The same applies to climate change—except that it is clear that the risks are so high that we need to accept the situation as real and take action on that basis.

Many of the voices of dissent about climate change come from, or are associated with, the financial and political sectors of our societies. Some business groups and financial interests do not want to acknowledge the problem because they think that by doing so they will acknowledge that they need to do something about it. The actions they will then be obliged to take might cost them money and reduce their profits. Or, possibly even worse in their view, they may be obliged to acknowledge the need for regulation of polluting activities and industries. Supporters of an unregulated free market are therefore very likely to look for reasons to reject the evidence for climate change. This is known as confirmation bias: we place greater weight on evidence that confirms our beliefs, while ignoring or resisting conflicting evidence. Politicians who deny the reality of climate change (frequently on the basis of very little information and large doses of self-interest) tend to do so because they serve electorates in which a large proportion of the population is itself ill-informed and does not accept the evidence for climate change.

Denialists frequently "cherry pick" evidence. They choose data that appear to contradict the body of evidence and assert that those data disprove the main conclusions. For example, they focus on short periods when surface temperatures bounce up and down from year to year for various reasons, such as heat exchange between the atmosphere and oceans. It's always possible to find short periods within long-term warming trends where temperatures fall briefly—this is just the nature of natural fluctuations about long-term averages in many natural systems. But the long-term trends are far more important than short-term fluctuations. Much is currently being made of the fact that the warming trend has slowed down over the last few years, although the absolute temperatures, relative to historical values, have consistently reached record highs.

There are even, bizarre as it may sound, those who argue that the whole thing is a vast conspiracy: that all those thousands of independent-minded scientists have perpetrated a huge "con trick" on the world and its populations, in the interests, we are told, of maintaining their funding. And so we could go on. It is a favorite ploy of right-wing talk-show hosts and denialist groups to assert that, because there are a few scientists who express doubts not only about the accuracy of various predictions but also about the reality of climate change, the scientific community is divided. This is blatantly untrue, and anyway, this is not a democratic discussion (if it was, the majority would win hands down); it's a discussion about facts and their implications. Furthermore, quite frequently, those who are loudest in their denialism have no background in climate science or ecology. We are not among that very small proportion of scientists who do not yet accept the evidence for climate change: we accept the reality of global warming and are concerned to explore its consequences. Denialism will not help to solve the problems arising from those consequences.

On Variation and Uncertainty

We cannot expect changes in climate to be smooth and progressive through time. It's a matter of common knowledge that weather varies from year to year and the data in figure 4.3 show this very clearly. The daily rotation of the earth and the constant changes in the distribution of the energy from the sun that warms it, interacting with the layer of gases that comprise the atmosphere, constitute a system that cannot be, and never has been, stable. Most air movement (wind) across the pressure gradients produced by differential regional warming is turbulent, and turbulence is a chaotic and unpredictable process, at least at small scales. So there is always uncertainty about what the seasons will bring to any region in any year. In general we can put boundaries on this uncertainty; the fluctuations in the values of the weather variables that together make up climate fall within predictable limits, but this is becoming more difficult as average global temperatures rise. Year-to-year differences, and the uncertainty about those differences, appear to be increasing; temperature records are broken and extreme events occur with greater frequency. For the most part, GCMs mimic the seasonal and annual variations in weather with remarkable accuracy. When they do not, it provides an opportunity to investigate the reasons and improve the models, which is done by finding an explanation for discrepancies, not by empirical adjustment.

Averages may not be very useful numbers when it comes to assessing local and seasonal weather patterns and their likely implications. Figure 4.3 shows average conditions for the earth as a whole, which themselves vary, but these averages mask even greater variations in weather, both in time and space. When we are concerned with particular regions we will usually have to examine the historical data, and the predictions, for those regions. Global climate models are now good enough to do this. They indicate that global annual precipitation is expected to increase steadily. The reason for this lies in the fact that as the temperature of the atmosphere increases, the air can hold increasing amounts of water vapor (see the discussion on this point in chapter 3). This leads to increased evaporation rates, particularly from oceans once covered with ice, and the resulting larger amounts of water vapor in the atmosphere contribute to the probability of higher precipitation.

The increases in precipitation are not, and will not be, uniformly distributed: we cannot yet accurately predict where precipitation will increase significantly[3] so we cannot, at this stage, assess the probable effects of these increases on forests and forest growth. Such predictions will have to include the effects of increased CO_2 concentrations on photosynthesis, and have to take account of the availability of the nutrients essential to tree growth, particularly nitrogen.

Somewhat ironically, global warming also has the opposite effect: warmer conditions are associated with more frequent occurrence of damaging droughts in some areas. For the United States, the projections are that northern areas are likely to become wetter and southern areas, particularly in the west, will become drier. Much of the western United States has recently experienced prolonged and severe drought. It's interesting to note a comment in an article in the *New York Times,* that Texas is experiencing a drought so serious that water supplies are becoming a problem affecting lifestyle as well as the state economy. According to the Texas Commission on Environmental Quality, "the already-dry western half of the state is expected to be particularly hard hit by climate change. . . ."[4] Yet state leaders still question the scientific consensus that humans are a major cause of climate change. A striking example of confirmation bias!

There are similar projections about precipitation distribution for Australia, where annual rainfall has increased in the northwest but decreased in the southwest and along and inland from the east coast, that is, in central Queensland and New South Wales. These latter regions have recently experienced years of serious drought, followed by unusually wet seasons. Western Australia has experienced, and continues to experience (at time of writing), severe and prolonged drought.

These weather patterns cannot be unequivocally attributed to climate change, but the associations with it are strong and the probability that there is a link is high. They are consistent with the predictions of GCMs. These projections imply that seasonal and annual variations in precipitation will get larger, so that short-term predictions become more difficult. Similarly, the increasing temperatures associated with global warming are not uniform anywhere across the globe.

Rising temperatures are the most important facet of climate change (figure 4.3 and the discussion above), hence the use of the term *global warming*, but as we commented earlier, average temperature is only an indicative number. Higher average temperatures may result from higher minimums, higher maximums, or some combination of both. An interesting study by Nicholls (2006), using the instrumental record and a GCM, documented changes in the Australian climate and the way they are associated with global trends. Nicholls showed that the mean maximum temperature has increased over most of Australia, except in the northwest and along the south coast of Western Australia, where the maximums have decreased slightly. Mean minimum (night-time) temperatures have increased over nearly all of the country except for some parts in the inland northwest, although the areas experiencing lower minimums vary between seasons. As well as providing information about the reasons for average temperature rises, Nicholls' paper illustrates the effects of global warming in terms of increasing variation. It's worth noting that the summer of 2012–13 was the hottest on record in Australia, by all measures.

Extreme Weather Events

Climate change is increasingly associated with extreme weather conditions. An increased frequency of heat waves is an obvious consequence of global warming. And drought, which we have already mentioned several times, is a form of extreme weather. It seems that increasing frequency and severity of droughts is associated with climate change, although the connection is more difficult to establish than that with heat waves.

Warmer sea surfaces in the southern Atlantic and the Caribbean during the hurricane season—late summer—cause increases in the amount of water evaporated from the ocean surfaces into the low-pressure systems that develop into storms. The additional water vapor adds to the energy in the system and will increase the intensity of the storm. Hurricanes, called cyclones or typhoons in the Pacific region, generate damaging winds and heavy rain. They can do immense damage as a result of storm surges and flooding. In tropical forests they result in large areas of blowdown, which, under natural conditions, are part of the normal lifecycle of the forests. Trees fall and open gaps that are rapidly colonized by herbaceous vegetation and, in due course, new large trees. If cyclones strike and destroy oil palm plantations the result is likely to be rapid erosion from relatively unprotected soil, and longer-term damage.

The probability of higher precipitation leads to higher likelihood of flooding in some areas. There are always floods somewhere in the world: in the last few years there have been devastating floods in parts of Europe, in the Philippines, in Australia and the United States. Because these are extreme events, like hurricanes, the statistics describing the frequency with which they occur are not yet definitive, but the evidence is growing that they are occurring more often than in the past—at least in relation to the period for which we have data—and that the number of very severe events is increasing. This is often expressed by statements to the effect that one in one-hundred-year events are occurring, and will occur, once in every twenty (or fewer!) years.

Impacts and Implications on Climate Change

Climate change has global impacts that do not—at least at first sight—appear to directly affect forests. Rising temperatures are causing rises in sea levels round the world as glaciers and the polar and Greenland icecaps slowly melt, adding huge volumes of freshwater to the oceans. Higher water temperatures also cause thermal expansion of the water, and hence a rise in sea level, which has serious long-term implications for millions of people in low-lying areas of the world. Melting glaciers are likely to affect the water supplies of many countries, particularly on the west coast of South America, where they are a major factor in determining river flows. The melting of the Himalayan glaciers, which feed the great rivers of Asia, could have incalculable effects on the flow regimes of hundreds of rivers, including the Ganges, the Yellow, the Yangtze, and the Mekong, affecting huge ecosystems and millions of people.

The Arctic is warming faster than any other region of the earth. This is thought to be caused by the fact that ice and snow reflect back into space a high proportion—up to 90 percent in the case of clean snow—of the short-wave solar energy that strikes these surfaces. As the Arctic ice disappears, there is a positive feedback effect as dark water and land absorb far greater amounts of energy, so that warming and melting are accelerated. International concern is rising as it becomes clear that the north polar ice cap, where the extent of summer ice has been reducing for the last twenty years, is melting faster than ever. We note, in passing, that boreal forest ecosystems are particularly vulnerable to climate change because the vegetation is not adapted to high (relative to the norm for the region) temperatures, or to summer droughts. Increased tree mortality has been observed in North American boreal forests.

Effects on Forests

Turning to direct effects on forests, in relation to changes in precipitation in general, we cannot yet predict accurately where precipitation will increase significantly so we cannot, at this stage, assess the probable effects of higher rainfall on forests and forest growth. Changes in average temperatures, and the occurrence of extreme temperatures are likely, over time, to bring about changes in the distribution of tree species. For example trees adapted to warmer conditions may become established higher up mountain ranges than at present, as the snow lines retreat.

We noted earlier that global warming is associated with the more frequent occurrence of damaging droughts in some areas. These droughts may result in tree death. Joyce and Running et al. (2013) note that rates of tree mortality in western (US) forests are well correlated with rising temperatures and increases in evaporative demand. There is a strong likelihood that the droughts experienced in recent years in the western United States and across much of Australia have exacerbated the effects of the rising temperatures associated with global warming.

Warmer conditions also affect the lifecycles of insects, and the way they interact with the plants they attack (or live with in synergistic relationships) so that species that may not, historically, have caused major damage may start doing so. Woods et al. (2010) compiled information on the influence of changing climate on the behavior of forest pests in British Columbia. They noted that "a hierarchy of thresholds must be passed before outbreaks can occur," and that climate change appears to facilitate the breaching of those thresholds. After that, widespread pest outbreaks, beyond the control of natural enemies, may occur. This has happened, to varying degrees, with mountain pine beetle, Douglas fir beetle, spruce leader weevil, and defoliating insects such as Western spruce budworm. In virtually all cases the pest outbreaks are associated with changes in the lifecycles of pests and with stressed forest stands. The current outbreaks of mountain pine beetle and Dothistroma needle blight (a fungal disease) are unprecedented in their severity and extent.

Bark beetles preferentially attack weakened trees and the drought and recent high temperatures across the western United States have provided these in huge numbers. Rising temperatures push up the altitude at which beetles can survive as well as allowing them to increase their number of breeding cycles and attack tree species previously outside their range. The result has been spectacular swathes of dead trees across the forests of many of the Rocky Mountain states: Colorado, Montana, Idaho, Arizona, and New Mexico, as well as British Columbia. This not only constitutes major disruption of forest ecosystems in itself, but also increases the risk of wildfire—dead trees make excellent fuel. And indeed there have been a number of serious wildfires in the western United States within the last few years.

It's possible, in principle, to manage, or at least mitigate, the effects of problems such as the bark beetle outbreaks. Nothing can be done about the droughts, or the impact of warmer conditions on pest lifecycles, but since bark beetles preferentially attack trees weakened by water shortage, thinning stands at risk to fewer trees, with lower stand leaf area index (LAI), will reduce the water use of the stands as a whole, in the same way as natural tree death makes more resources available for the survivors. In the transition to different climatic conditions, and increases in disturbance by fire, insects, and diseases, we may expect shifts in forest composition and density. Some tree species will be more susceptible than others to these changes and a new mix in composition is expected, with or without the invasion of nonnative organisms to a region.

The climatic changes and drought in Western Australia are contributing to deteriorating woodland and forest health. During the record hot and dry period in 2010 and 2011, large patches of trees throughout the region suddenly died, with little recovery in some areas. In several ecosystems, species have died out and not been replaced, permanently shifting vegetation structure and ecosystem function (Matusick et al. 2012). These changes, like those in the drought, insect, and fire-affected forests in the United States, illustrate what happens when the resilience of a system reaches a tipping point and a new transition state occurs in a sudden, catastrophic step.

Elsewhere in Australia in 2010, catastrophic fires swept through the eucalyptus forests of Victoria, destroying huge areas of forest as well as small towns, where they killed more than 170 people. Widespread intense fires occurred again in this area in 2013, as well as in Tasmania. Concern has been expressed that these tall forests, which have survived for thousands of years, and are able to tolerate periodic low-intensity fires, may be destroyed by these high-intensity and increasingly frequent fires. The fires have followed periods of unusually high temperatures and droughts, making the forests tinder-dry and highly flammable. The situation is exacerbated by ill-advised policies that try to keep all fires out of the forests. This results in the buildup of high quantities of fuel—dead leaves, bark, and branches, the normal litter shed by the trees—so that, when there is a fire it is intense enough to scorch and kill mature trees as well as young, regenerating trees.

An obvious general point about forest wildfires is that they emit into the atmosphere some of the carbon stored in the aboveground parts of the trees. This is an acceleration of the natural process of decay that takes place when trees, or parts of them, die. Forests, particularly old forests where there are many large trees, store considerable amounts of carbon, but they are not necessarily dynamic sinks for carbon; that is, the rate at which CO_2 is absorbed and carbon stored may not always exceed the rate at which CO_2 is emitted, as a result of the decomposition of dead wood, foliage, and soil organic carbon. After a forest is burned, and particularly if salvage logging is then carried out, it will be years before it stores enough carbon to make up for the amount emitted in the fire, so if it's burned again before that point, it is a net source of CO_2 to the atmosphere. If the resilience of the forest ecosystem is perturbed past the tipping point, the forest may be replaced with some other, less desirable, ecosystem.

Young, actively growing forests, on the other hand, will be positive sinks for carbon—they will be sequestering it rapidly. This is why plantation forestry is frequently proposed as one of the ways of combating anthropogenic carbon emissions, although the proponents are not always aware (apparently) that as the stands approach maturity, growth rates fall, litterfall rates increase, and the carbon balance approaches neutrality. Furthermore, most plantations end up as pulp and then paper, which has a very short "residence time" before being burned or consigned to landfill. In either case the carbon is rapidly returned to the atmosphere.

Annually, forests capture and store, at least temporarily, around 20 percent of the carbon dioxide that modern human activities emit into the atmosphere; without them, climate change would be accelerating even faster than it is now. In later chapters we deal with the carbon balance of forests, which underlies the whole question of forest growth rates and has to be understood if forests are to be managed effectively.

Summary

Carbon dioxide (CO_2) concentrations in the earth's atmosphere are increasing as a result of human activity. This is reducing the amount of long-wave radiation that escapes from the earth, which alters the energy balance of the earth and is causing global temperatures to rise faster than at any time in the past. Other gases (methane, nitrous oxide) are also involved. The primary sources of CO_2 emissions are fossil fuel burning and land clearance; when land, particularly forested land, is cleared, the carbon stored in the plants is released to the atmosphere as CO_2.

The rate at which the earth is warming is clearly demonstrated by temperature records taken around the world (the instrumental record), and is confirmed by models based on the physics of the general circulation of the atmosphere and its interaction with land and water surfaces. These models are well validated and provide the tools to make predictions about future climates, given assumptions about CO_2 emissions. There is always variation in weather patterns, but the long-term global warming trend is consistent.

Some people—mainly politicians and business people with special interests such as coal and oil supplies—deny that climate change is really taking place, denying the reality of evidence and cherry picking data, such as short-term fluctuations, to argue that the situation is "normal." This will not help solve the problem.

Global climate change is associated with increased probabilities of severe drought in parts of the world such as the western United States, and Western Australia, and with increasing numbers of extreme weather events, such as hurricanes and serious floods.

The effects of climate changes on forests include the direct effects of damaging droughts and also possible changes in the distribution of tree species. The probability of intense and dangerous wildfires increases in drought-affected forests. Indirect effects, through changes to the lifecycle of insects that attack trees, particularly those weakened by adverse conditions, may be serious: there are examples of this in the United States and Canada. Some diseases may also be favored by rising temperatures.

Chapter 5

How We Value and Use Forests

How we value forests, and the importance that we attach to conserving and managing them depend, to a large extent, on our own personal philosophies and backgrounds. Our values, and those of others, are inextricably associated with the intended uses of forests.

People who live in and depend on forests are likely to think about the way they are used very differently from those who live in the cities of developed Western countries. We pointed out in chapter 1 that there are still people in the tropical forests of the Amazon, Indonesia, and Papua New Guinea who depend on the forests they live in for the materials to build their houses, for most of their food—obtained by hunting and gathering—and for plants with medicinal properties. The number of such people is now small, and there are probably few, if any, who have not been exposed to the modern world, its products, pressures, and way of life. The pressures on them and their environment are, in many cases, very strong. Logging and mining companies push into the forests, sometimes with the approval of national governments and sometimes illegally. Small groups of indigenous people who, in most cases, hold no formal title to their land and who are unskilled in political negotiation, are generally helpless to resist encroachment and the destruction of their way of life. We can't speak for them but simply note that their rights must be considered: their continued existence provides a strong argument (among many) for the preservation of large areas of tropical forests.

Those who manage forested land for wood production are likely to take a fairly utilitarian attitude toward it: their objective is an economic return from the land. This is not to say that they don't value it from an aesthetic or ecological point of view, or that management for wood production necessarily results in serious reduction in other values, but it may do so. We discuss the management of forests used primarily as wood-producing systems in chapter 6, where we consider the economic factors that underlie management practices and decisions.

A great many people value forests mainly for their aesthetic values—their contribution to the beauty of scenery, or as places to walk and appreciate nature—or for ethical reasons, based on their appreciation of the ecological services provided by forests, from which everyone benefits. The attitudes of people with this point of view are likely to be different from those of production foresters, although there is no fundamental reason why the objectives of the two groups can't be reconciled.

Conservationists, hydrologists, and climatologists are concerned with the role and importance of forests in local, regional, and global ecosystems. They think in terms of the processes described and discussed in chapter 3, and the ecosystem services forests provide. These include the absorption of carbon dioxide (CO_2) and carbon storage, clean air, clean and reliable water supplies, and the conservation of biodiversity. The knowledge generated by those who study these matters is important for ensuring that the yields from forests managed for wood production are sustainable,[1] ideally, even under changing climatic and economic conditions

Since we are forest ecologists, our perspective in relation to the use of forests is strongly influenced by our knowledge of the ecosystem services they provide. That value system underlies the following discussion. However, we recognize the importance and economic value of forests as wood-producing entities, and we are not arguing that ecosystem services and ethical and environmental value systems are necessarily better than, or should outweigh, the conventional economic considerations that underlie wood production operations. Such judgments are, to a large extent, philosophical matters and outside the scope of this book: readers will have their own positions. We observe, however, that if unsustainable forest practices and policies are followed, societies suffer.

Ecosystem Services and Universal Values of Forests

The word *use* in the title of this chapter implies active utilization. But, besides the products derived from active utilization, forests provide ecosystem services that support the web of life and benefit humanity as a whole. Three are of particular importance: the reliable maintenance and supply of clean water; the absorption of CO_2 and storage of large amounts of carbon in forest biomass, and the maintenance of the biodiversity. All are essential for the long-term health and resilience of the world's plant and animal populations, and therefore for the health of human populations.

Reliable Supplies of Clean Water

The principles underlying the flow of water from catchments—the water yield—are contained in equation 1 in chapter 3, and some of the implications of clearing forests from these areas are discussed in general terms in that chapter. Obviously, the flow of water from a catchment depends not only on the vegetation cover (e.g., forest, grassland, agricultural crops), soil type, area, and topographic features, but also on the frequency, intensity, and physical nature of the precipitation. Vegetation cover is seldom homogenous, particularly in areas that have been developed for human occupation, so comparisons between watersheds are invariably difficult, but we can make some valid general statements.

Peak outflows usually occur during and after rain events large enough to generate maximum stream flow from runoff. For comparable sized catchments and amounts of precipitation, these will always be lower from one that is forested than from those where the vegetation is predominantly grassland or crops. (Of course if large areas are covered by houses and paved roads, almost all rainfall or melting snow will run off immediately.) The high runoff rates from cleared land are likely to produce soil erosion. This is always a problem in cropping systems, particularly those that leave the soil unprotected by vegetation for significant periods of each year. Rapid runoff causing erosion carries soil that causes dams to fill with silt. It also carries contaminants, such as pesticides, and nutrients that degrade water quality and can encourage algal growth in water storages. Forests will not stop floods resulting from exceptional precipitation, but their root systems hold the soil and provide insurance against washouts and landslides of the sort that occur frequently in high rainfall areas denuded of forest cover.

Base flow is the term used to describe longer-term rates of water supply from a drainage basin. Base flow is the result of lateral movement of water through saturated subsurface soil. It depends on the amount of water that has infiltrated into the soil, and on soil properties. Because infiltration rates are high, the proportion of the water in a given precipitation event that runs off across the surface of forest soils is generally small. As a result the amount of water stored in forest soils tends to be high, and base flows are maintained for long periods.

There is a great deal of research, reported in the technical literature, which indicates that, in general, the water yield of forested drainage basins is more stable than that from cropped or rangeland catchments. However, water yields are often lower from forested catchments because forests tend to transpire more than other vegetation types. Their canopies also intercept more water than smaller vegetation. This water is lost by evaporation. In general, water yield increases following reductions in vegetative cover. Research shows that thinning forests increases the water yield of catchments and causes groundwater levels to rise. This is because thinning reduces the leaf area index (LAI; see chapter 3), so that both transpiration and interception of rain or snow by the canopy are reduced. In parts of Australia (and no doubt elsewhere), the removal of forests and woodlands has resulted in rises in groundwater levels, which has had the unwanted effect of bringing toward the surface salt stored deep in the soil. This leads to rising salinity levels in streams and the surface soil, and has adverse effects on crops in the affected areas.

The possible benefits of the higher water yields that may be obtained by thinning or removing forests, which civic authorities concerned with water supplies to cities might see as desirable, may be more than offset by possible losses in the stability of water yield, and by poor water quality. A study by the US National Research Council[2] on the hydrologic effects of forests showed that forests provide natural filtration and storage systems that process nearly two-thirds of the water supply in the United States and that healthy forest vegetation can benefit human water supplies by controlling water yield, peak flows, base flows, sediment levels, water chemistry, and quality. Forest cover in catchments can substantially reduce the need for treatment to treat water for domestic use, and thus significantly reduce the costs of supplying water.

A 2008 report from the UN Food and Agriculture Organization (FAO)[3] supports many of the points we have made: forested watersheds provide high-quality water by minimizing erosion, reducing sediment movement into water bodies (dams and lakes), and by trapping and filtering other contaminants and pollutants.

Carbon Dioxide Absorption and Storage

The process of photosynthesis by which forest canopies absorb CO_2 was outlined in chapter 3, and some of the consequences of deforestation and land-use change were discussed in chapter 4 in relation to climate change. Since the absorption and storage of carbon (C) by forests is inarguably an important ecosystem service, we present, in table 5.1 data for the fluxes of C into forests, emis-

TABLE 5.1. (a) Carbon (C) fluxes and (b) storage in the world's forests

(a)

Forest type	Boreal	Temperate	Tropical intact	Tropical regrowth	Tropical deforesta-tion	Tropical land-use change
Fluxes (Pg C yr^{-1})	0.5±0.08	0.72±0.08	1.18±0.41	1.64±0.52	−2.93±0.47	−1.29±0.70

(b)

Forest component	Live biomass	Dead wood	Litter	Soil
Stored C (Pg)	363±28	73±6	43±3	383±30

Note: 1 Pg = 10^{15} gm = 10^9 tons. Carbon in CO_2 is 27% of the total mass, so a ton of carbon is equivalent to 3.67 tons of CO_2 absorbed, emitted, or stored by forests, i.e., 1 unit C ≈ 3.67 units CO_2.

sions from them, and the amounts stored in different forest types. Fluxes are positive values, and emissions, negative values. The uncertainties in the estimates are the values following the ± operator. Storage data describe the total amount of C stored globally in different components. The data in this table are derived from Pan et al. (2011). Carbon in CO_2 is 27 percent of the total mass, so a ton of carbon is equivalent to 3.67 tons of carbon dioxide absorbed, emitted, or stored by forests.

The data in table 5.1 indicate that, globally, forests absorb much more carbon than they emit. However, the very large negative fluxes, that is, C emissions, from tropical deforestation and tropical land-use changes, outweigh, by a wide margin, the absorption of carbon by those systems, illustrating the contribution that deforestation in the tropics makes to the accumulation of CO_2 in the earth's atmosphere (see also figure 4.2). It's also important to note that the amount of carbon stored in forest soils is slightly larger than the amount stored in live biomass. The authors of the paper from which these data were taken found that there is a fundamental difference between tropical and boreal forests in their carbon storage patterns: tropical forests have 56 percent of carbon stored in biomass and 32 percent in soil, whereas boreal forests have only 20 percent in biomass and 60 percent in soil. The relatively small proportion of carbon stored by tropical forest soils is a result of the rapid decomposition and turnover of organic matter in those soils, while the high storage in boreal forest soils lends emphasis to the point made in chapter 4: the area supporting boreal forests contains a large proportion of the world's soil carbon. If the decomposition of this carbon is accelerated by rising temperatures, it will release enormous quantities of CO_2 into the atmosphere.

Biodiversity

Biodiversity refers to species richness: the range of plants and animals, which includes insects and microorganisms that exist in a specified geographical region. The word may also be applied to genetic diversity.

Biodiversity is central to ecosystem functioning. Healthy ecosystems contain a range of animal, bird, and insect species that fulfill various functions: predators and prey, pollinators and seed disseminators—some of which are species-specific, some general. Both live and dead organic matter is consumed and broken down by a multitude of organisms. A healthy ecosystem is dynamic; there is continual change and interaction among its components. Different plant species fill various niches in ecosystems by exploiting environmental resources (light, water, nutrients from the soil) in different ways, reacting differently to changing conditions. They are eaten by different insects and other animals, and seeds are disseminated by a range of mechanisms. Communities (ecosystems) with high biodiversity are far more resistant to serious damage by a disease, or suites of diseases, that may attack some species but to which others are resistant. They are also likely to contain species with a range of abilities to withstand changes in environmental conditions such as those likely to result from climate change. High biodiversity usually ensures that ecosystems are both resistant to change, and resilient in the face of disturbances.

Genetic diversity means there is a wide range of genes within plant communities. In natural ecosystems there is genetic diversity within and among different species. In monocultures, if there are genetic differences between individuals within the population, some are likely to be resistant to attacks by insects, or to infection by disease organisms. This ensures that, when attacks or epidemics occur there will be survivors that can pass on their resistance to future generations. Clonal monocultures, where all individuals in the population have the same suite of genes, can be susceptible to devastating damage if attacked by a disease or insect species to which the genotype is susceptible. The genetic diversity in forests must be maintained to provide the genes needed as the basis for breeding programs to produce trees with desirable wood production properties, and with the ability to resist or recover from disease and insect attacks. The latter are likely to become more frequent and severe in future (see the report by Woods et al. [2010] on forest health and climate change discussed in chapter 4). The genetic resources of forests, particularly tropical rainforests, also provide a valuable reservoir that humans may need to draw on, for example, for the genes we need to restore the resilience of our agricultural systems or to provide the basis for novel pharmaceutical drugs. Humanity's current dependence on the productivity of vast monocultural systems, in turn dependent on massive chemical inputs, is dangerous and inherently unstable.

Some plant species have medicinal properties: quinine, the first chemical (we would now call it a pharmaceutical) to be effective in the treatment of malaria, comes from the forests of South America. People indigenous to the tropical rainforests are known to have identified thousands of plant species that have biological effects—medicinal, psychological, or recreational—on humans. Shamans have been experimenting with various combinations and dosages for generations. Rainforest plants are the source of many of the drugs, both pharmaceutical and recreational, used in the developed world.

Since there are still thousands of plants that have not been examined for possible medicinal properties, it's safe to assume that there will be, among them, a great many that provide chemical compounds with medicinal value. The US National Cancer Institute estimates that 70 percent of the plants with anticancer properties are found only in tropical rainforests (see Butler 2012; also, on the *National Geographic* website, "Rain Forest: Incubators for Life"[4]).

The Uses of Forests

Humans have, from time immemorial, lived in and used forests, and we have already mentioned that there are still groups of people for whom the tropical forests provide everything they need in their lives. We also noted in chapter 1 that in the developed countries—northern Europe and North America, but also Australia—people use the forests for recreation: for hiking, camping, and hunting and, particularly in the Scandinavian countries, for berry and mushroom collecting. The social importance of these activities varies from place to place. Our concern, in this chapter, is with the major direct use of forests for wood production, a use and activity that has been important for thousands of years. We discuss the modern situation.

Harvested Wood and Its Uses

Wood products can, for convenience, be considered in three major categories: pulpwood, sawn timber, and wood for fuel.

Pulpwood

Pulpwood is used predominantly to make paper, cardboard, and packaging materials. Despite the decline of newspapers as the modern world moves toward electronic communications, there is still a vast market for newsprint. Consumption of newsprint is declining in the Western countries, but in China and India there are still hundreds of high-circulation newspapers, consuming thousands of tons of newsprint daily. Add to that magazines, advertising material (including vast quantities of "junk mail"), wallpaper, wrapping paper, office, or copy, paper, and toilet tissue of various types and it seems clear that the markets for pulpwood are likely to remain strong, at least for the foreseeable future.

Stands harvested for pulpwood are invariably *clearcut* (also referred to as *clear-felling*). The trees are cut, debarked in the field or at the mill, and pulverized either mechanically or with chemicals to separate the cellulose from the lignin fibers. The chemical process is known as the *Kraft process*. It's the more common one in modern pulp and paper production, producing pulp that can be used to make higher-quality papers than the mechanical process. The wood from different tree species has different properties in relation to pulp for papermaking, the most important of which are differences in length of the wood fibers and in wood density[5]—the number of fibers per unit mass. In general, both dense hardwoods, such as oak, beech, and eucalyptus, and lighter ones, such as poplar and willow, have shorter fibers than the softwoods, which include pine, spruce, and fir. The pulp from different species can be mixed to give the papermakers more control over the quality of paper they produce.

Until about the 1990s, most pulpwood came from naturally or artificially regenerated forests, that is, forests that had been previously logged and allowed to regrow, in Canada, the United States, and northern Europe, with a significant proportion from the softwood boreal forests. In recent years, the proportion of pulpwood from plantations has been increasing steadily, and the bulk of pulp production now comes from plantations. In fact, a 2007 FAO report[6] shows that over 50 percent of pulpwood traded on international markets now comes from eucalypt plantations. Much of the expansion in plantations is in tropical areas, particularly Latin America and the Caribbean, where the forest products industry is largely based on plantations. Nevertheless, Europe, including Russia, is still the largest paper products exporting area. The extensive softwood plantations in Chile, New Zealand, and South Africa provide both pulpwood and sawn timber products. The industries in Chile and New Zealand, and the wood chip industry in Australia, are strongly export-based, but because these are small countries their impact on the overall world markets is also small.

It's interesting to note from the FAO report (2007) that paper and cardboard consumption—reflecting pulpwood consumption—has fallen slightly in North America, reflecting the global financial downturn in 2008, but it is rising steadily in Asia, reflecting the rapidly expanding Chinese economy.

WOOD FOR FUEL

For much of human history wood was virtually the only fuel for cooking, heating, and processes such as smelting (see comment in chapter 1). The Industrial Revolution of the nineteenth century, originating in Britain, was driven by steam but would not have been possible without coal, which became the major fuel source. Coal was also used as domestic fuel, and the smoke produced by countless coal-burning stoves and heaters was largely responsible for the appalling smogs that were characteristic of industrial Britain in the nineteenth and early twentieth centuries. But although wood had been supplanted as the major industrial fuel, it remained important in most parts of the world, not least in the United States where, although coal provided the energy for most industry in the nineteenth century, it was wood that fired the boilers of the iconic steamboats. And, as the railways pushed across the continent, their boilers were also, initially, wood fired. The demand for wood by the steamboats and railways resulted in the destruction of huge areas of forest, particularly along the Mississippi, where the consequent bank erosion contributed to the increasing silt load of the river system.

In today's world, wood remains an enormously important fuel, with the advantage that it is a renewable source of energy. The FAO report (2007) shows that, of the approximately 3.5 billion cubic meters of timber harvested in the world in 2007, slightly more than half was used for fuel.[7] Much of this is in developing countries with large populations of poor people; the rural poor in particular have no options but to use firewood for cooking. Their need to collect wood for this purpose involves endless, tedious labor, the burden of which falls primarily on women. It also results in denudation of woodlands and, in some cases, forests. Another little-recognized effect of firewood collection is the removal of all litter from plantations, short-circuiting the recycling process and contributing to the progressive loss of nutrients from the system. Soil structure also tends to degenerate as a result of the progressive loss of organic matter.

These problems—fuelwood availability, soil degradation, and nutrient depletion—can be solved by establishing plantations specifically for the purpose of providing domestic fuel. Australian researchers have been active in this area, investigating the burning properties of wood from various species. The ubiquitous eucalypts are now widely grown specifically for fuelwood, but some other species, notably *Acacia mangium* have been shown to serve the purpose well.

In the developed world, technological developments have made wood a far more efficient fuel in terms of the energy released by combustion. This is achieved by producing low moisture-content pellets and briquettes, burnt in customized stoves and boilers, where the combustion process is usually accelerated by fans blowing air into the combustion compartment. The technology has been developed to the point that wood pellets are now used to provide the heating, ducted from central burners, for apartment blocks. This use of these wood products has an energy cost, in producing the pellets or briquettes, but major advantages in ease of transport (the products are much lighter than "raw" wood), energy release, and reduced pollution because of the clean combustion. Pellets and briquettes can be produced from wood pulp or the waste left after logging operations; there are now a number of mobile pellet-producing machines available, and no doubt there will be continued development in this area.

There is research, in a number of places, on the production of *ethanol* from wood. Ethanol is the form of alcohol used to supplement petroleum fuels. It's currently produced mainly from sugar cane (Brazil is the leader in that) and from corn, in the United States. Production from corn distorts the market for the grain as a food source and, in a time when food production looms as an increasingly important problem for the world's rapidly increasing human population, it is, arguably, a serious misuse of resources. The ethanol produced from corn is called starch ethanol. Starch is a relatively simple carbohydrate and the process of converting starch into sugar and then fermenting that to produce ethanol is straightforward. Ethanol from wood is called cellulosic ethanol. *Cellulose* is far more complex than starch, and the ethanol production process is more complicated and currently too expensive for widespread use.

There are claims that economically feasible techniques of ethanol production from wood are being developed. If true, we may expect that, in the near future, corn will be replaced as the feedstock for the process by low-quality pulpwood and wood waste from forestry operations. Such a development will be particularly important in Europe, which has neither the corn production potential nor the fossil fuel reserves available to the United States. However, as with most technologies, there is a potential downside: if the debris left by forest operations was to be removed consistently, there would be progressive reductions in the organic matter content of forest soils, leading to long-term deleterious effects on fertility and soil stability. The likelihood of this happening would be increased if tree stumps and root boles were removed.

Charcoal—essentially carbonized wood—has a long history. In fact much of the wood used as fuel in processes such as smelting, before coal displaced it, was in the form of charcoal. There are various methods of making it: the basic process is a controlled, slow, cold burn that dries and carbonizes the wood. Traditionally, in many countries, including much of Europe, the charcoal burners were people who often lived isolated lives in the forests, exercising their inherited skills and experience to produce a valuable fuel. Charcoal remains an important fuel today. Inevitably, there are industrial methods of making it.

SAWN AND PROCESSED TIMBER

Sawn timber comes from trees that have good form, that is, they are large and straight enough to provide good quality planks and boards. For hundreds of years, in many places throughout the world, houses were built entirely from timber, and this is still the case in many places. Even where houses are built mainly from brick, many of them contain internal walls based on timber framing, and roof trusses are generally timber. House construction remains one of the largest consumers of solid timber; periods of economic downturn and slow house construction have direct effects on the markets of solid timber.

There are now a number of substitutes for timber from large trees. Small wood can be debarked, shredded, and glued to produce various products such as particle, chip, wafer, fiber, or strand board. These can be made to any required dimensions; they are workable, like wood, and can be produced with great structural strength. Such products provide another alternative use for smaller wood pieces.

The other major use for sawn wood is furniture manufacture. China is currently the world's largest exporter of wood furniture, using wood produced domestically and imported. China is also a major importer of tropical hardwoods. Because the market it provides for this timber is largely unregulated, it is indirectly responsible for much of the illegal logging that is contributing to the destruction of the forests of Southeast Asia.

In the past, when trees were felled using axes and the timber taken out of the forests by horse- or ox-drawn wheeled vehicles, so-called selective logging was the normal practice. Generally, relatively few of the trees in a stand were removed, leaving the others to take advantage of the gaps created and become larger, perhaps to be logged in another cycle if they were of high enough quality. Nowadays, the advent of highly effective chainsaws and powerful, heavy machinery has made it possible to cut trees rapidly and move large quantities of timber quickly. Clearcutting has become the normal practice. Logging roads are driven into the forests by bulldozers, and everything in a stand is cut down and delimbed; bunched and loaded by grapple hooks, cranes, or forklifts; then trucked to a mill or exported. In some operations in old forests where most of the trees are large, all the marketable trees will be harvested and the tops and branches left as debris on the site. In other cases, the large trees are taken to sawmills, and the smaller are converted to wood chips, which then go for pulp. We consider the implications of clearcutting natural forests, and the management of logging debris, in chapter 6.

In Australia there have been long-running controversies about the logging of old-growth forests. Most of the state forest services were established in the late nineteenth and early twentieth centuries, with the primary objective of ensuring that supplies of timber were adequate to meet the requirements of the developing economy. Management to conserve flora and fauna was always a secondary objective which, in most of the states, was later handed to departments of wildlife and conservation. Since there was virtually no *silviculture* involved, the practice was inexpensive, and the foresters argued that it was sustainable in terms of wood yield—the primary criterion by which the operations were judged. However, there was growing unease among increasing numbers of people in the wider community about the ecological consequences of clearcutting native forests, particularly mature, multistoried (so-called old-growth) forests.

The essence of the argument runs like this: the foresters asserted that the plant communities that emerge from the seedbeds created by clearcutting and burning the *slash* are new and vigorous versions of the old communities. This is difficult to support with empirical evidence: there are likely to be species changes, and if the clearcut is planned as the precursor of another logging operation in, say, forty to eighty years (a commonly cited rotation length for saw log production), then by definition the forest is not the same, nor will it store the amount of carbon present in the original stand. The operation might be sustainable in terms of wood yield, but it is not sustainable in terms of equity between generations: subsequent generations will not be able to enjoy the aesthetic benefits and ecosystem services of great old trees, and some of the more specialized plants and animals that reside in such habitats will not survive.

There were economic as well as ideological reasons for the controversy. Because of the history of forestry in Australia, the logging companies, throughout the forested areas, were equipped to deal with large trees, as was the sawmilling industry. In response to pressure to move their operations into plantations, the loggers and sawmill owners claimed, with justification, that there was not enough wood available from plantations—certainly not enough of saw-log size and quality—to support the industry and provide the country's sawn timber requirements. It was also claimed that relatively young, plantation-grown wood was not suitable for sawmilling (the technical problems in this respect have now been largely overcome). Furthermore, sawmills would have to retool to deal with smaller trees. Unfortunately, the resistance within the industry to change its practices translated into a reluctance to invest in the establishment and management of plantations, so that the transition of the forestry industries from native forests to plantations has been slow. However, the areas under plantations are increasing, and the wider community and ecological considerations have prevailed, if not yet entirely.

The controversy (which has many more complex ramifications than we have been able to outline here) encapsulates the difficulty of changing entrenched positions and adapting to the needs of a rapidly changing world. It has now been largely resolved, but the political compromises involved in the resolution mean that mature native forests are still being logged. There was a similar argument in New Zealand where, from the time of colonial settlement, there was massive forest clearance for agriculture as well as logging for sawn timber. It was resolved by heavy investment in plantations and government decrees that stopped all logging in native forests.

Abuses of Forests

If we accept that the preservation of remaining forests should be an important objective for any country, for reasons relating to the ecosystem services they provide and the values discussed in the previous sections, then we should accept the assertion that any human action that directly results in the destruction or degradation of forests can be called abuse. This proposition may seem to run counter to the requirements of increasing numbers of humans for land for agriculture and living space, both of which result in forest destruction. It would be inappropriate in this book to try to explore the implications and arguments in particular situations, but we note, in passing, that the solution to the impending problems of feeding the world's human populations should not lie in trying to clear more tropical forests for agriculture. An analysis by Foley et al. (2011) led to the conclusion that tremendous progress in meeting the world's food requirements could be made by halting agricultural expansion, closing "yield gaps" on underperforming lands, increasing cropping efficiency, shifting toward more plant-based diets, and reducing waste. Together, these strategies could double food production while greatly reducing the environmental impacts of agriculture.

Urban Development

Most of the abuse of forests is in the tropics, but growing populations exert pressure everywhere. In California, for example, although society there is very aware of environmental issues, the demands for a pleasant lifestyle in beautiful surroundings have led to houses being built in forested areas. These bring with them access roads and the disintegration of forests into separated subunits within and between which the integrity of the forests is compromised. Urban development of this sort also brings with it pressure on authorities to prevent fires, resulting in fuel buildups and serious problems during drought periods in hot summers. The damaging fires experienced in California in recent years are largely a consequence of this sort of development, exacerbated by the droughts associated with climate change and the excessive accumulation of fuel.

Australia has the same problem. We noted earlier that the fires in the state of Victoria in 2010 destroyed several small towns and killed 170 people. Much of the damage was directly attributable to the fact that the towns were deliberately built in the forests—they were lifestyle choice towns; people wanted the sylvan environment, the trees and the wildlife, beauty and peace. All good, except that they also demanded that the forests should be as undisturbed as possible—there were even regulations against collecting dead wood for firewood. So when extreme conditions happened—dry, very high temperatures (>40°C, i.e. >104°F) and strong winds— ignition occurred from power lines brought down by wind, and the result was a massive firestorm. Such high-intensity fires are destructive not only to humans but also to the forests themselves, where the lack of fuel management led to accumulations of highly combustible material, and the fires killed trees that would have survived moderate burns.

Tropical Forest Destruction

Much of the extensive logging that takes place in tropical rainforests is illegal, particularly in Indonesia, Laos, Cambodia, Thailand, Papua New Guinea, and Brazil (although that country is making serious efforts to control it). It is driven by the market for tropical hardwoods in China and Korea.

Papua New Guinea's (PNG) forests are the third largest, and some of the most diverse, on earth. However, the World Bank estimates that over 60 percent of PNG's original forests have already been destroyed by logging and industrial agriculture, and that 70 percent of logging in PNG is illegal. According to the environmental organization Greenpeace,[8] illegal logging costs timber-producing countries between US$10 and US$15 billion per year in lost revenue. This accounts for over a tenth of the worldwide timber trade, estimated to be worth more than US$150 billion a year. But Greenpeace also points out that importing countries must share the blame for the devastation. China, with its current focus on development and economic growth, is unlikely to be interested in doing so.

In Asia the pressures on forests are from burgeoning human populations, as well as from ruthless commercial logging subject to very little control in most countries of the region, and from clearing for crops such as palm oil. Such clearance not only destroys the forest (which is the intention when plantations are to be established) but where the slash (foliage, branches, tree tops) left on the ground is burned, it creates serious smoke pollution over the region. At the time of this writing (mid-2013), the blanket of smoke from illegal land clearance in Indonesia, drifting over Malaysia and Singapore, is dense enough to constitute a health hazard and cause international protest. And this is a regular occurrence.

The palm oil industry in Indonesia and Malaysia is developing rapidly. In 2012 the Indonesian government designated about 35 million hectares of already-logged forest for farmland or exotic tree plantations. The sequence is this: the forests are logged and damaged, then said to be too badly degraded to be worth preserving, so they are made available for "development." There is a website[9] where the ecological implications of the oil palm plantations are documented. They include destruction of orangutan populations, which are in danger of extinction; the destruction of ecosystem values and services; and a range of adverse effects on indigenous human populations. Many oil palm plantations are being established in low-lying areas that were rainforests, where deep peat has accumulated in the swampy conditions. Forest clearance and burning lead to the decomposition of this peat and massive emissions of CO_2.

We have already commented on the dangers and short-term benefits of farming on land that was tropical forest: lacking the organic matter additions from litterfall and rotting material, tropical soils become depauperate within a few years, and will not produce worthwhile crops. Oil palm plantations have similar effects. They do not provide significant protection against runoff and, within ten to fifteen years, the soil nutrients have been oxidized, volatilized, or leached out.

There is extensive land clearance for agriculture in the Amazon, as well as damage to forests caused by mining and river damming. Damage to and exploitation of the forests there are driven by ever-growing human populations and the ever-growing demand in international markets for hardwood timber. In South America, roads are being pushed into the forests and the rural poor move in along those roads; according to Bill Laurance,[10] 95 percent of all deforestation in the Amazon occurs within 10 km of roads. Peasants move in along these roads and carry out slash and burn, shifting agriculture. These people can't be condemned for their actions, even though those actions are frequently illegal. They usually have very limited opportunities to make a decent living, but their agricultural practices destroy the forests. In Africa, too, the greatest pressures on the tropical forests come from expanding populations and constant clearance for agriculture. We have provided some additional comments on the reasons for forest destruction in the humid tropics in chapter 7.

Summary

The values placed on forests by humans vary. Those who live in forests and depend on them do not see them in the same way as those who live in developed countries, who may value forests for aesthetic reasons. The livelihood of the more forest-dependent people is under threat from logging and clearing forests for agriculture (which includes plantations).

Forests provide ecosystem services of universal value. These include the absorption and storage of carbon dioxide, reliable supplies of clean water, and biodiversity in the flora and fauna populations. The value of wood products is more easily appreciated by most people. Logging practices to obtain these products vary from careful selective logging to clearfelling (clearcutting), which is now the normal procedure in many places. Wood is used for pulpwood, sawn timber products for building and furniture making, and for fuel. Wood used for fuel may be burned directly, or it may be converted into charcoal or wood pellets and briquettes to feed fires. The use of wood as a major source of fuel on railways and riverboats was a significant factor in the destruction of large areas of forest in the United States. It is likely that ethanol production from wood will become economically feasible in the near future.

Urban development, particularly in the western United States and in Australia, is steadily expanding into forested areas. This causes fragmentation and, because fuel builds up in forests around urban developments, fire management is difficult. When fires occur they are damaging and dangerous. Large areas of temperate forests are being destroyed for urban development in western North America as well as in smaller areas in Australia. Even larger areas of tropical forests are lost to clearing and burning in Indonesia and other Southeast Asian countries, as well as in the Amazon basin.

Chapter 6

The Economics and Practices of Forest Management

This chapter provides an outline of the economic principles and ideas that can and should be applied in the management of forests for wood production. This type of business involves decisions about planting, weed control, fertilizers, pesticides, thinning, harvesting, and the use and control of fire, all of which incur costs and have economic consequences. Wood provides economic returns, but it's not easy to put prices on some of the forest values we discussed in the previous chapter.

Management involves making decisions, deciding what to do in situations where alternative courses of action are available; implementing the decisions; and evaluating the results. All the decisions in forest management—as in the management of any enterprise or activity that involves the use or manipulation of physical, biological, or financial resources—involve assessing the likely costs of actions in relation to the expected returns. In classical (or, as it's sometimes called, neoclassical) economics, costs and returns are evaluated simply in terms of money, and the implications of decisions can, for the most part, be evaluated in financial terms.

This is true for forests when we consider only (or primarily) the production of wood and wood products, commodities for which there is a market. But, not all decisions can be evaluated in these terms. For reasons we discuss a little later, much of the theory that underlies neoclassical economics can't be applied in forestry.

Economics is often broken down into *microeconomics*—usually regarded as the economics of households and small businesses—and *macroeconomics*, concerned with large-scale economic factors such as employment and national productivity. We will discuss macroeconomics (very briefly) in terms of *gross domestic product* (GDP) and *natural capital*—the value of the forests and their soil, of water quality and biodiversity. We consider microeconomics as economic factors relating to particular forests, or the activities of forest companies. In relation to microeconomics, there are important differences between state or publicly owned forests and privately owned forests. The factors that influence the decisions that have to be made by the managers of companies operating in natural forests on state-owned land are different from those that influence the decisions made by managers of companies operating in plantations grown on private land.

The most basic relationship in classical economics is the *supply and demand law*: if there is an oversupply of a commodity in the market, the price will tend to fall; if there is not enough available to satisfy demand, it will tend to rise. But economic systems are a great deal more complicated than that. They include the materials, capital equipment, and skilled labor needed for production, which in turn are heavily influenced by technology. Other factors in economic systems are the distribution networks, the financial system and the way it is affected by modern communications, the value of money and international exchange rates, the demand by human populations for particular products, the stimulation of demand by advertising, and so on. Basic economic theory relies heavily on assumptions about perfect and rational markets, within which the players have perfect knowledge and in which there are no externalities to distort production and consumption decisions. An externality exists whenever the true cost of producing something, such as wood, is not reflected in its selling price. Externalities also exist when individuals are not prepared to pay for a product or service, or claim not to value it, although they cannot be excluded from the benefits of the service. Carbon absorption and storage by forests is usually treated as an externality, as are natural capital values.

The assumptions underlying classical economic theory bear little relation to the real world and the effects on markets of human preferences, foibles, and sometimes irrational behavior—not to mention knowledge that is always far from perfect. Furthermore, macroeconomics is as much a matter of politics and sociology as it is of the exchange of goods, services, and money. The idea of perfect knowledge is even more unrealistic in relation to forestry than in most other areas of human activity. (It's worth noting that the erratic and sometimes wild fluctuations that characterize stock market prices, which periodically lead to bubbles, booms, and crashes, can't be predicted or explained in terms of supply and demand, competition, or so-called business cycles.)

In relation to the economics of forestry, there is also the question of response time. Those who operate in money and investment markets can respond almost immediately to changes in market conditions (which is one of the reasons that stock exchanges are so volatile), and most manufacturing industries can respond quite rapidly to changes in market demand. However, in the case of forestry and forest products, even for operations in natural forests, it may take several years for operators to "gear up" to increase the amount of wood they can harvest (assuming the wood is available) in response to increasing demand. And, if demand and prices fall, they will have difficult decisions to make about whether to stop operations, which would mean that machinery representing a large and depreciating capital investment would stand idle, and workers may have to be laid off. Alternatively, they can continue to operate at a loss. The gamble managers have to take depends on a great many factors, including their "reading" of the market and its future. In the case of companies based on privately owned plantations, the decisions are likely to be even more difficult. They have to decide whether to maintain planting programs to produce trees that will not be ready for harvest for anything between one and ten decades in the future (depending on location and species). Classical economic theory has little to offer people who must make decisions with such timeframes. We will come back to these points later, when we discuss discounting.

The idea of market failure is pervasive in economics. Economists have quite precise ideas about how perfect markets might look and operate, but they are acutely conscious that unregulated markets frequently produce less than optimal outcomes. There are many examples in modern life. The major causes of market failure include structural problems such as monopoly and oligopoly,[1] information asymmetries where buyer and seller have different information about demand for, or availability of, a product, and the existence of externalities, which are especially important in forest management, particularly at the macroeconomic level.

The reason for this lies in the values of forests that we discussed in chapter 5: the ecosystem services provided by forests tend to be treated as uncosted externalities. Utility companies that provide clean water charge for it, but it's very difficult to put a price on the contribution that the forest cover in watersheds makes to the quality of that water. It's also difficult to put a monetary value on the absorption of CO_2 by forests and on other ecosystem services, such as biodiversity, that they provide. So these contributions are usually ignored in economic calculations, with the result that the natural capital embodied in forests is frequently degraded because it's cheaper to use the forests that way. We discuss this in a little more detail later. When ecosystem services are not taken into account in making decisions about forested land, the risk that the forests will be degraded increases.

One of the most important objectives of forest managers must be to maintain sustainability. This means that the capacity of the forest to provide goods and services should not be permanently diminished over time by any of the operations carried out in the forest. The

Canadian Forest Service defines sustainability as "Management that maintains and enhances the long-term health of forest ecosystems for the benefit of all living things while providing environmental, economic, social and cultural opportunities for present and future generations."[2] We might rephrase this by saying that we believe natural forest ecosystems should be maintained, as far as possible, in a series of states that are not greatly different from those that exist under natural (nondegraded) conditions, so they can provide to future generations the products and services that are enjoyed by the current population.

Maintaining sustainability doesn't mean that forests should be left untouched. There must be reserves where preservation of biodiversity is the management policy, but most natural forests can be used for wood production without damaging them seriously. There is always natural change, but disturbances caused by human actions, such as selective harvesting, need not degrade the forests beyond the points at which they lose their stability as ecological systems (see the discussion of resilience in chapter 1) and flip to some new, degraded state from which they can't recover. (We recognize, in chapter 7, that transitional states may be the new norm.) This prescription can be relaxed in the case of plantations, although soil stability and fertility still have to be maintained in those systems.

An Outline of Relevant Economic Theory

This section deals with the ideas and principles underlying economic analysis and micro- and macroscale economic decision making. Microscale refers to microeconomics, defined earlier as concerned with economics at the level of individual properties or companies; macroscale refers to macroeconomics, concerned with the economics of states and countries. Our use of scale should not be confused with its meaning in discussions about "*economies of scale*." In the context of production systems, increasing the scale of an operation leads to fuller utilization of resources because more products can be manufactured with the same investment in plant and equipment. Economies of scale are also achieved in operations such as strip mining, where investment in more massive machines improves output efficiency. They apply to some extent in forestry operations, where decisions may be made about whether to expand the production capacity of an enterprise, for example, by buying more land, but we won't be going into all that here. In the section on forest growth and yield estimates, we outline how the amount of wood in forests, and rates of wood production, are estimated.

Macroeconomics as It Relates to Forestry

We live in a finite world with finite resources. The conventional measure of economic growth used in modern societies—called gross domestic product—attempts to capture the total market value of all the goods and services produced in a country (salaries and wages paid, purchases of goods and services, mortgages and rents, income from trade, etc.). Economic growth is measured in terms of increases in the value of GDP from one period to another. GDP is often assumed to be a good indicator of the well-being of societies—although, in fact, it isn't. It measures all sorts of things that add to economic activity but are nonproductive and likely to detract from human happiness and welfare.[3]

More important, in relation to this discussion, is the fact that the depreciation of capital that accompanies production—in particular natural capital—is not deducted from GDP. Dasgupta (2010) has provided a cogent discussion about the consequences of this. He says "We economists see nature, if we see it at all, as a backdrop from which resources and services can be drawn in isolation. Accounting for nature, if it comes into the calculus at all, is usually as an afterthought to the real business of 'doing economics.'" Where forests are being degraded or destroyed in pursuit of economic growth, the result is depreciation of capital assets. No economist or businessperson would regard such a practice as a sensible, justifiable, or sustainable way of running a business. So why do we behave that way in relation to natural resources?

Dasgupta (2010) argues that the reason is the failure to secure property rights to natural capital. This includes communal property rights (say, to the forests in which people live) and global property rights, which would be pertinent in discussions about climate change and what to do about it. (Who "owns" the problem?) For most ecological resources there is no market because, in many cases, relevant interactions with economic consequences take place over large distances, making the cost of negotiation too high. For example, upland deforestation may lead to erosion that affects downstream water quality and activities such as fishing. Silting also degrades estuaries, which are often important fish breeding grounds. But the people downstream are unable, in most cases, to bring any pressure on those harvesting the forests, so the downstream problems are treated as uncosted externalities by the harvesters. If a country gives timber concessions to private firms (as happens in many Southeast Asian countries and in Canada), the result might be an increase in revenue to that country from royalties and taxes on exports. But there's not much chance the local people affected by the logging (this doesn't apply in Canada) will be compensated for damage and possible destruction of the environment they depend on. Even in countries with economically mature economies, it's difficult to enforce on the logging companies laws intended to provide safeguards against serious environmental damage. The result is that those extracting the logs, who do not pay recompense to those who bear the costs—either present or future generations—are subsidized by nature and by the country from which they are extracting the resources. Where natural resources are underpriced, they are overexploited. If the natural assets of a country are permanently degraded, the wealth of that country will decline.

In general, forest products appear in national accounts simply as entries in the budget papers, noting (for example) the value of wood or wood products sold and imported. The importance attached to forestry policy is likely to depend on the importance of the timber industry to the national GDP. The constraints imposed by government on operations in state-owned forests are likely to reflect political compromises between the demands of industry, which usually wants to operate with as few constraints as possible, and those who argue for the preservation of natural capital and ecosystem services. The *stumpage price* (royalties) charged for wood extraction from those forests will usually be negotiated between the state and industry. The state derives benefits from its forests through the value of the ecosystem services they provide, but stumpage provides revenue and is supposed to reflect the value that society puts on the wood to be harvested. Whether it reflects the true value of the public goods that may be destroyed in the process of logging is likely to depend on the political flavor of the government and its perception of the importance of forests and forest values, and on the power and influence of the timber industry.

Forest Management Policies in Different Countries

Obviously, the legislation governing industry access to public forests, and the constraints placed on industrial operations, vary from country to country. We outline here, very cursorily, the situation in a few countries to illustrate the range of factors that affect the economics of forest management and the decisions forest managers have to make. It becomes clear from these outlines that there is considerable overlap between the macro- and microeconomic levels, mainly because most forest operations (except fire control) in publicly owned forests are carried out by private timber companies. The economics of their activities are determined by, and can be analyzed in terms of, "standard" neoclassical principles.

UNITED STATES

Forest policy in the United States centers around the management of national forests by the US Forest Service. Few areas are leased long term to timber companies. The Forest Service has adhered, since the 1990s, to the definition of sustainable forest management given earlier for the Canadian Forest Service. Many of the states, starting in the 1970s with Oregon, began to legislate forest practices that apply to both private and state forestlands. The federal forests agreed to meet or exceed these increasingly strict standards. In California, 15 percent of all state and private ownership must include old-growth native forests. In California, Oregon, and Washington, clearcuts are restricted to < 50 ha, and buffer strips of uncut forests are required along all fish-bearing streams. Additional restrictions are placed on logging in areas with endangered species such as the spotted owl.

Since the first forest practice laws were passed, there have been regulations requiring that successful regeneration be established within a few years of cutting. Roads must be built with landings situated to minimize erosion and surface runoff. More recently, when roads are repaired, culverts must be replaced to ensure access to streams by fish. On federal land large areas are set aside as reserves where, as in some national parks, prescribed fires (and untended wildfires) are allowed, with the intention of fostering conditions thought to be more typical of historical norms (although current and predicted changes in climate are not considered). This may be influenced by the fact that nearly 50 percent of the federal agency's budget now goes to controlling fires.

Although forest growth rates now exceed the rates of cutting in most national forests, the maintenance of biodiversity remains a concern because the landscape is highly fragmented, partly because of restrictions on the size of logging units and the interval mandated for cutting adjacent units. Some new policies are being considered that would prevent clearcutting while allowing frequent access to produce gaps in mature, multistoried forests by thinning. There is a requirement for speedy replanting of native stock, but this has had the unintended consequence of limiting plant communities that normally establish themselves after events such as fire or windstorm. As a result, browse for ungulates, and food and cover for many other species, are reduced. But the impact of climate change is now widely recognized, in relation to reestablishment policies.

AUSTRALIA

In Australia the primary responsibility for forest management rests with state (as opposed to the federal) governments, and the basic thrust of forestry legislation was, until recently, concerned with the exploitation of timber resources. The aim was to provide timber for the construction industry, other economic uses, and wood chipping, as well as to provide jobs for timber workers. Recent legislation, driven by increasingly well informed public opinion, has moved toward the incorporation of multiple use and ecosystem values.

There is still conflict about logging native forests, particularly in Tasmania, where the forestry industry remains an important factor in the economy of the state. Much of the argument turns on the questions of what "ecologically sustainable" means: does it mean, as we said earlier, that natural forest ecosystems should be maintained in states that are not greatly different from those that exist under undisturbed conditions? Or does it mean, as the forestry industry has argued, a healthy, often even-aged forest, regenerated after clearcutting? The arguments have also hinged on the question of how much wood is in the forests. It has been agreed that specified areas will be set aside to ensure the survival of the local industry and the economies of timber towns. But it's surprisingly difficult to come up with figures for the amount of wood in a heterogeneous old-growth forest, as well as estimates of the regrowth rates that determine rotation time (the period between harvests). Therefore, it is difficult to reach agreement on the area of forest that would have to be set aside for the wood production needed to support already-established industrial logging, and the jobs that go with that. We will come back to this question of yield estimation later in this chapter.

The plantation industry has been encouraged by various taxation measures and some subsidies. However, misleading advertising about potential growth rates and plantation productivity and, hence, returns on investments, has led to widespread economic losses to stockholders and subsequent destruction of many plantations that had inadequate market value.

CANADA

In Canada, forestry and the timber industry are major contributors to the national economy, and the environmental/ecosystem services viewpoint seems to be a major factor influencing forestry legislation. Canada has been committed, since 1992, to sustainable forest management with the aim of maintaining a vibrant forest economy while protecting the health of forested lands and maximizing the environmental and social benefits from those lands.[4] Some of the same problems mentioned for the United States in regard to fragmentation of the landscape apply to Canada. Also, in both areas where primary (previously uncut) forests are logged, the removal of standing and fallen dead trees impoverishes the system's ability to maintain soil organic matter and diversity. On private lands, application of chemicals, while strictly regulated, still creates concerns about their effect on biological diversity and water quality.

BOREAL FORESTS IN SWEDEN AND RUSSIA

A paper by Elbakidze et al. (2013),[5] about the implementation of forestry policies in Sweden and Russia, illustrates the influence of policy and legislation on forest management. About half of Sweden's forests are owned by nonindustrial private forest owners, and the rest mainly by industrial forest companies, the state, and forest commons (land in common ownership by farmers). The Swedish Forestry Act regulates rotation periods and the regeneration practices that must be followed after final felling of stands. It also specifies the ecological, social, and cultural values that must be taken into account. The policy gives equal priority to production and environmental objectives, and aims at ensuring sustainable production. Only about 1.3 percent of the total forested area is cut each year, the clearcuts being followed by planting or leaving seed trees for regeneration on suitable sites.

In Russia, virtually all the forests are owned by the state. They are leased by private forest companies and management is regulated by a forest code, which defines the upper limit for timber harvest in terms of an annual allowable cut over particular regions. In the area that Elbakidze et al. (2013) studied, ecosystem values are maintained by setting aside forest strips along rivers and streams to protect water quality and soil, and along fish spawning areas. Areas with high biodiversity are also protected by national nature conservation and forest legislation. The maximum allowable cut is estimated from standing wood volume inventories and rather poor models of forest productivity. Furthermore, in calculating the allowable cut, the areas set aside to conserve ecosystem values are not excluded, so the allowable cut from the wood-producing area is overestimated. The result is "forest mining" by clearcutting. Regeneration is left to nature. Large regions of the Russian boreal forests were severely affected by accelerated logging during recent decades and the intact forest landscapes are shrinking by 3 percent annually due to these practices.

TROPICAL RAINFORESTS

International concern about the destruction of large areas of the rainforests in the Brazilian Amazon has been instrumental in causing the government to tighten regulations about forest logging and the management of those parts of the forests where logging takes place. It has also moved to reduce illegal logging and forest destruction. There are projects aimed at ensuring that management practices maintain the forests as sustainable wood producing and ecological systems. There is still a long way to go, but progress is being made. Brazil is also a major producer of plantation-grown wood.

We mentioned in chapter 5 the forest destruction that is taking place in Indonesia. There are international efforts to halt it but, unfortunately, there is little evidence of progress, and the forests in Indonesia are to a large extent exploited for commercial gain, rather than managed. This country provides a prime example of the exploitation of natural capital, resulting in its depreciation.

Those concerned with the situation in any particular country will find it necessary to get detailed information about the forest industry and its practices in that country. These are always influenced by the economics of forestry in the context of the national accounts, and by the presence and strength of environmental movements and activists. Scientific ecological principles are seldom the first priority of governments, or politicians in general, but if the world's surviving forests are to be managed sustainably we have to keep propounding those principles.

Microeconomics: Commercial Operations on State Land

Table 6.1 provides an outline of the costs that will be incurred by timber companies who work in publicly owned natural forests. That means they work on land they do not own and for which they do not carry the capital costs. We have not included, in this table, the costs of finance that companies may have to meet. These will vary depending on a range of factors such as ownership, age of the company and how it was set up, and whether it carries major debts. The financial returns come from the sale of the harvested wood. They will depend on the market for that wood and the quantity sold.

The actual costs associated with any particular item will

TABLE 6.1. Costs and returns likely to be incurred by a timber company working in state-owned forests

Costs	Returns
Stumpage	
Access roads	
Machinery: harvest	Wood sales
Machinery: transport	Sale of pulpwood, saw logs, biofuel
Labor	
Replanting and weeding?	

obviously vary enormously, and not just with the economic demand for products. The stumpage charge will depend on forest type and will be based on estimates of how much wood is in the forest as well as the quality of the wood—high-grade saw logs and pulpwood will be valued differently. Stumpage will not, normally, include costs of land husbandry—management for environmental values, and the maintenance of natural capital. Estimation of the volume of sound timber in a stand is central to the economics of stand management; we provide an outline of how this is done in the section "Forest Growth and Yield Estimates" in this chapter. Wood yield estimates can be highly variable and quite inaccurate. We noted earlier that stumpage is negotiated with the state and it may, but usually does not, include allowance for ecosystem values.

Operational costs will vary depending on whether the stand is to be clearcut or whether selective logging for saw logs or (electrical) utility poles or mine props is to be carried out. Terrain will affect the cost of logging, dictating the cost of roads, or alternative ways of extracting forest products (cable logging, helicopters, with horses or elephants).

Clearcutting in even-aged stands of moderate-sized trees (say, diameters not greater than about 30 cm) is usually done, nowadays, with harvesting machines that grab the trees, cut them near the base, strip off branches, and cut the stems into designated lengths. The cut logs are stacked by mobile cranes and loaded for transport to the mills. Such machines now have massive low-pressure tires to minimize soil compaction. They are almost universally used in plantations. If the stands are old-growth, and logging is selective, the trees will be cut manually with chainsaws. This is the usual procedure in tropical forests, where the target is high-value large trees. If the procedure—including log removal—is carried out carefully the damage to the forest is minimal and logging is sustainable. But the way logging is being carried out in many tropical forests—using heavy-tracked equipment; driving bulldozed roads into the forest with no regard for stability, drainage, or plants; and burning the residues from the operations—is destructive and unsustainable.

Stands being managed for the production of saw logs may be thinned at some point in the rotation. *Thinning* consists of removing a proportion of the trees in a stand so that the remaining trees will grow faster and end up larger. Decisions about whether, and when, to thin, involve silvicultural judgments about the likely benefits of the practice and financial judgments about whether it will be worth it. The silvicultural options are myriad: it is possible to thin a stand several times, the first time perhaps precommercially, that is, cutting out trees too small to market, with the benefit of reducing competition in the stand. Later thinning will usually only be done if there is a market for the wood. When it is done, and how much is taken out, will vary with species, stem populations, growing conditions, and other factors that managers will have to assess. If the wood can be sold at high enough prices to offset the costs of the operation, then it is more likely to be done than if there is no market. In the latter case, stands will probably only be thinned if, in the view of the forest managers, based on their experience and calculations of current and projected growth rates, the gain in growth of the trees left standing is likely to be greater and yield wood of more value than the cost of the operation. Discounting procedures (described later) should be used to make these assessments.

Early thinnings may be sold for pulpwood and the stem populations reduced to quite low levels. The surviving trees are then left to grow, possibly for many more years, to produce large stems suitable for high-quality sawn timber. Quality can be improved by pruning lower branches back to the stems shortly after the last thinning, to reduce the knots in the wood. The operation may be repeated as the crown "lifts," that is, as the trees get taller and the physiologically active canopy moves higher. Thinning is carried out in national forests in the United States to improve the vigor of stands and make them more resistant to insect attack. Because it reduces the *leaf area index* (LAI) it will also reduce water use, at least in heavily thinned stands, provided understory vegetation is also controlled.

We have put a query against the "replant" entry in table 6.1, since in many cases this will not be required and will not be done.

Regeneration will then be left to nature. The extent to which companies are obliged to ensure that their operations do not degrade the forests (as in Canada and Sweden, discussed earlier) is likely to be strongly dependent on the legislative constraints imposed on them by the government authorities. It will affect the costs of their operations.

We see no point in trying to put indicative figures on the entries in table 6.1; they are extremely variable. The managers responsible for such operations will need to know the numbers and how they vary. For any particular forest the stumpage costs will be fixed. Others, such as operating costs—which depend on conditions— machinery maintenance, and depreciation, will be variable. Transport costs will vary enormously, depending on forest type, topography, the existence of roads within the forests and the costs of their maintenance and/or construction. Distances to sawmills or pulpmills have to be taken into account. The wages paid to workers vary from country to country. Labor costs are also affected by factors such as whether there are enough workers available locally and whether they must be transported to the job, or provided with accommodation, all of which vary from country to country and place to place. All this would have to be taken into account in any detailed economic analysis of particular forest operations. The manager's job is to keep costs as low as possible so that profits are maximized in relation to the returns expected from the wood.

One of the conventional ideas in neoclassical economic theory is cost-benefit analysis. This, as the name implies, is a procedure for working out whether the economic benefits likely to accrue from a particular purchase or practice are likely to outweigh the costs. Such analysis might be applied to things like the purchase of new machines, hiring more labor, or opening up new areas of forest. It always involves assumptions and estimates, not the least of which is "what price are we going to get for the wood?" In the case of an operation where the wood is standing in the forest and the current price is known, if there is only a short interval between incurring a cost, such as buying a new machine, and the improved production benefits that can be expected from that, the analysis is relatively simple. However, things get a lot more complicated when the machine has to be amortized over a number of years. Wood prices become more uncertain the longer the period involved, so the long-term profitability of the machine becomes increasingly uncertain. The question of long-term expectations about prices and returns on investments leads us to consider discounting, which we discuss briefly in the next section.

In contrast to the operations of timber companies in countries such as Canada, Australia, the United States, and Europe, small-scale (village-based) commercial forestry in state forests would, in principle, be a good option for indigenous people in places such as Indonesia, Papua New Guinea, the Pacific islands, Thailand, Cambodia, tropical Africa, and South America. Given support to set up sawmills, local timber industries using careful selective logging techniques could provide incomes to small communities, and their activities could undoubtedly be indefinitely sustainable. The problem is political will and corruption. Large timber companies with their destructive methods make large amounts of money from tropical hardwoods. They can pay bribes in the states in which they operate, to ensure that those operations are unhampered by oversight or constraints of any sort.

Microeconomics: Commercial Operations on Private Land

We consider, in this section, commercial forestry operations on land owned by companies or, more rarely, by individuals. This will, in most cases, be plantation forestry and the economics of management will be very different from those that apply to state-owned land. Many of the points we make in this section apply as much to saw logs as to pulpwood production, but there are large differences in the size range of sawmills and pulpmills as well as in wood quality requirements. To keep things reasonably simple, we will confine this discussion to pulpwood production.

The first major cost is the land itself. If it has to be purchased, this will involve a major capital outlay, in most cases using funds provided by banks or finance houses of some sort. Such funds come at a cost, either in terms of interest payments and some arrangement for capital repayment, or because the financing agency has an interest in the business and therefore requires regular dividend payments. These factors come into play when companies or individuals intend to buy land for the primary purpose of wood production. If the land is already owned, and wood production is to be a new venture, the financing costs will depend on the kind of contract the landowner has with the pulp producer who will buy their wood. Or landowners may simply allow a company that supplies timber to sawmills or pulpmills to harvest standing timber on their land, in which case stumpage charges will be negotiated between the owners and the harvesting company.

We assume that the landowner(s), whether individuals or companies, have done their market research homework. Individuals and smaller companies proposing to produce pulpwood will need a contract with a mill within a reasonable distance. Some large companies may construct their own paper and pulpmill. To operate economically mills must have year-round guaranteed supplies of timber; that is, they require even wood flow. This requirement may be met by contract arrangements between the mill owners and private growers, which can include assistance in establishing plantations. In Tasmania the owners of a large pulpmill contracted to plant, maintain, and harvest plantations on private land; in other words they carried all the costs of producing the wood and took responsibility for wood flow. Landowners were simply paid for the use of the land. However, in other (probably more usual) circumstances, landowners would have to meet the production costs.

Clearly there is a large range of possible scenarios relating to land ownership and wood supply, but to illustrate some aspects of the economics that underlie management decisions and actions, we will model (in an elementary way) a hypothetical situation

TABLE 6.2. Economic considerations governing wood production operations on private land, aimed at serving a pulpmill

	Symbol	Example data
Total area under plantations (ha)	A	100,000
Harvest rate (ha day^{-1})	h	50
Av. yield at harvest (m^3 ha^{-1})	y	300
Harvest operations (days yr^{-1})	d	320
Av. rotation length (yrs)	R	6
Replant rate (ha day^{-1})	r	50

where a large company is embarking on an operation to provide pulpwood to a mill. We assume the company owns an estate of 50,000 hectares suitable for plantations and intends to use all of it. We refer to table 6.2.

If the operation aimed at and achieved a planting rate of 2,000 ha yr^{-1} (i.e., about 8 ha day^{-1} for a 250-day working year) then it would take twenty-five years to plant the whole estate. If the planned rotation age is fifteen years, then the first harvests will be in year sixteen, so in that year, and subsequently, harvest operations will have to be carried out simultaneously with new planting and replanting harvested areas. Until the first harvest there would be no income from the land. If the average yield at harvest is expected to be 300 m^3 ha^{-1}, then assuming wood density is 0.5 kg m^{-3} the harvest will yield (300 x 0.5) x 2,000 = 300,000 tons pulpwood per year. Income will be dependent on the price paid for the pulpwood.

The company, in its planning, would allow for increased labor and greatly increased machinery and transport requirements once harvesting started. This company would, presumably, have a contract with a pulpmill to supply specified amounts of wood from the date when they expect to start harvesting. From the point of view of the mill manager it is essential that even and predictable wood flows be maintained. Pulpwood plantations are seldom thinned.

Replanting costs include the costs of a plant nursery and the complex logistics of propagation and ensuring the supply of enough seedlings, of appropriate size for planting out, to meet replanting requirements. In some eucalyptus species, new stems grow from the stump of harvested stems; this is known as coppicing. It usually produces a bush of young stems, which have to be manually pruned to one when they are established, but the cost of doing this is far lower than the cost of replanting. When using coppicing species it may not be necessary to replant more than once every two to three rotations. On the other hand, genetic selection of better adapted species or varieties may justify replanting.

The scenario outlined above could apply to eucalyptus plantations in Australia or South Africa, *Pinus radiata* (Monterey pine) plantations in New Zealand, or loblolly pine (*P. taeda*) in southern parts of the United States. It is obviously highly simplified, but it serves to indicate some of the complexities involved in setting up and managing a large wood-production operation. (Detailed analysis would have to take account of land preparation, road making and maintenance, transport and labor costs, etc.) It also allows some exploration of the implications of changes and variation: for example, if average yields are not being achieved in some units, then it may be necessary to increase the area harvested to maintain supplies to the mill. Any of the operations outlined—seedling production, planting, harvesting and transport—might be contracted out or done by the company. Various alternatives can be explored by using different values for the data in the table.

In any realistic planning exercise of this type the management team would undoubtedly use computers with digitized maps of the plantation estate, containing for each management unit, soil type, planting date, possibly historical yield, and so on. They may also use a decision support system to enable them to assess various options and alternative strategies to ensure even wood supply and to evaluate the economic returns from different options. Estimation of the projected yield from any unit is an essential part of the financial and logistical planning; what this involves is described later. However, before we move to that we must consider the question of discounting, a financial technique essential to economic planning in relation to forestry.

Discounting and Net Present Value

In standard neoclassical economics production is assumed to be more or less instantaneous, followed immediately by sales, so costs and returns are assumed to occur in the same timeframe. However, by investing money in something that's expected to provide a return in the future, we forgo the benefits (consumption) that we could have had by spending the money now. A monetary investment is expected to yield a real rate of return, but inflation and constant decreases in the value of money are realities of our age, so the returns expected from funds committed to an investment are likely to be lower, in real future terms, than their apparent current value. So future values must be discounted to provide estimates of net values in terms of what the funds would be worth now. That means that, when you calculate the returns you expect from an investment committed for some specified period, you should discount the value of the expected return by an assumed inflation rate and fall in the value of money. In the context of forestry, the value of the expected return depends on the expected yield, which must be predicted (we discuss, in the next section, how this is done) as well as the expected value of the product.

For example, you invest $1,000 for ten years at a fixed compound interest rate of 4 percent. Using the standard equation, and expressing the interest rate as a fraction, that is, 0.04:

Future value = Present value x $(1 + \text{interest rate})^n$
that is, $FV = PV(1+r)^n$, where n is the number of years

to calculate the returns, gives ~ $1,480 at the end of the investment period. But if we assume that the inflation rate will be 3 percent, then to estimate what the future value of our investment would be in current monetary terms, we have to discount it by that rate. We do this simply by rearranging the equation to give the *net present value* (NPV):

$$NPV = FV/(1+r)^n \sim \$1,101.$$

If the estimate of inflation that is used is wrong in either direction, the estimate of NPV will also be wrong. If inflation exactly matched the interest rate earned then NPV would equal PV.

From the point of view of investments in growing wood, this kind of calculation has to be applied to the expected returns on wood harvested. If we take the situation outlined in table 6.2, there must obviously be quite massive initial investment, which we will ignore, as well as ongoing costs of land preparation, planting and replanting (or coppice management), harvesting, and transport. The first returns from the enterprise will come when the first trees are harvested, fifteen years (in our example) after planting. Let's assume that the company guesses that the price of pulpwood in fifteen years is $500 per ton. On that basis, at the end of the first harvest year the company can expect to realize $(300,000 \times 500) = \$150 \times 10^6$. However, discounting the return at 2.5 percent (see table 6.3) would reduce the $500 to a NPV of $345 over fifteen years, so the return would be $(300,000 \times 345) = \$103.6 \times 10^6$. A more pessimistic estimate of 4 percent for the discount rate reduces the $500 to $277.60 and the total NPV to $\$832 \times 10^5$.

A company doing calculations of this type would, of course, examine a number of possible scenarios. If the price of pulp rose faster than the cash depreciation rate they used in their discounting calculations, then the forward estimates would indicate that the future value would be higher than the PV, and vice versa. Also, as time went on and the first harvests approached, estimates of yield, pulpwood price, and the appropriate discount rate should improve steadily, so the calculations would become increasingly accurate.

TABLE 6.3. Net Present Values of expected returns on pulpwood production after growing periods (rotations) of 6, 15, and 25 years, assuming discount rates of 2.5% and 4%

Pulpwood prices	$500		$750	
Discount rate	0.025	0.04	0.025	0.04
Rotation length (years)				
6	$431	$395	$647	$592
15	$345	$277	$518	$416
25	$270	$187	$404	$281

Table 6.3 shows the results of NPV calculations assuming future pulpwood prices of $500 and $750 per ton applied to rotation periods of 6, 15, and 25 years. Six-year rotations are achieved in Brazil and Venezuela, using eucalyptus. Fifteen years is more typical of *Pinus radiata* (Monterey pine) plantations in New Zealand or eucalyptus plantations in Australia. (Most Australian eucalyptus plantations are in the cooler southern parts of the country, where growth rates are much slower than in subtropical South America.) A 15-year rotation is also typical of densely planted, unthinned plantations of loblolly pine (*P. taeda*) in southern parts of the United States, although rotations twice that long, combined with thinning (and with the sale of hunting permits), can bring in much larger net returns.

Discounting calculations would be made by any plantation grower needing to work out the optimum time to cut his or her trees. Tree crops have the advantage that, if the market falls the crop can be left standing and will not lose (intrinsic) value; in fact, the value may appreciate as the trees will continue to grow. So if the current pulpwood price is only $350 ton^{-1}, a plantation with estimated standing biomass of 250 m^{-3} ha^{-1} (which, since the density of wood is about 0.5 tons m^{-3}, yields about 125 tons of wood) would be worth \$43,750 ha^{-1}. However, the owner may decide to wait for a few years before harvesting and selling the wood. If he estimates the discount rate at 3 percent, and delays harvest by two years, the NPV would be $350/(1+0.03)^2 = \$330$ ton^{-1} if the price remained the same. To improve on his expected income the price would have to go up to at least $350(1+0.03)^2 = \$371$ ton^{-1}. The delay may also provide some savings in harvesting and reestablishment costs.

Decisions about discount rates depend on the way people read the markets, including the international situation with regard to trends in wood and paper use. In fact there is a great deal of guesswork involved, and the normal procedure would be to run a number of scenarios with various assumptions about forest growth rates and yields, which depend on factors like growing conditions and likelihood of disease or insect attack, as well as various financial considerations. Estimates of future income streams also provide managers with the information they need to calculate the net profits of their operations: they have more control over costs than over market prices and inflation rates.

Forest Growth and Yield Estimates

Estimating the amount of wood in a management unit, now or at some future date, is an essential prerequisite to any decision relating to the economics of forest management. It's also a necessary process for making estimates of wood supply for the purposes of budgeting and estimating future income, as well as for estimating harvesting and transport costs. The procedures and techniques for making these estimates—called collectively, forest modeling—lie at the heart of most forestry and forest management courses. Empirical forest models are based on data obtained from statistical analysis of measurements and stem counts, collectively called *mensuration* or *biometrics*. There are numerous books and innumerable technical papers dealing with mensuration and forest modeling; all we will do here is provide an outline of some of the main approaches used, with some assessment of their strengths and limitations. We also provide a short description of process-based models and the progress being made in the use of remote sensing to estimate forest biomass over large areas.

Empirical Growth and Yield Modeling

The basic relationship underlying growth and yield modeling is as follows: yield = (number of trees on a site) x (tree size). Conventionally tree size is expressed as wood volume, in cubic meters per hectare.[6] Since mass is (volume x density), where density is expressed in kg m^{-3}—in the case of wood this refers to the density of dry wood—the mass of wood in a stand can be directly estimated from volume measurements.

Much of the complexity of forest modeling arises from the fact that, even in plantations grown from genetically uniform clonal material, shapes of individual stems vary slightly. In natural forests containing a range of species they vary a great deal. However, we can start with the assumption that the trees in a stand are of similar size and taper uniformly from the base to the top of the stem. In this case, we can calculate the volume (V) of wood in a stand from measurements of stem diameter (d_B) and height (h), using the standard equation for the volume of a cone, that is, $(^1/_3 \cdot \Pi [d_B/2]^2 \cdot h)N$, where N is the number of trees per unit area (hectare).

Of course, things are not that simple. In any stand where there is significant variation in the sizes of trees, we will have to deal with size classes, which means deciding on size-class intervals. Even in uniform plantations, trees are not likely to be perfectly conical, and the top part of the stem is not usually of much value, even as pulpwood. So resources inventory, as it is called, involves measuring the heights and diameters of a sample of trees in a stand, and developing statistical relationships between stem diameters and tree heights and volumes. This usually involves destructive sampling— cutting down trees to make detailed measurements of stem mass and volume on the ground. Diameters are conventionally measured at a so-called breast height, now standardized to 1.37 m (4.5 ft.). Stem diameters and tree numbers per hectare (*stocking*) provide estimates of the basal area (BA) per hectare. There are several optical instruments available for measuring tree heights, but the procedure is complicated by visibility problems and by slope.

To deal with these problems measurements are usually made in standard diameter sample plots, chosen as far as possible to be representative of the whole area of interest. The number and size of the plots necessary to provide good estimates of wood volume will vary depending on the uniformity of the site and of the stand. Variability in the size and shape of the trees in natural forests is likely to be much greater than in plantations. In the case of large plantation estates or natural forest areas it may be necessary to establish sample measurement plots in all the units across the estate that are on different soil types and possibly subject to varying rainfall. The diameters and heights of all the trees in these sample plots are measured and the resulting estimates of wood volume are scaled up to the whole forest.

Another procedure used is variable radius plot sampling, which involves estimating tree sizes, and hence their basal area, using an angle gauge prism. The number of trees with stem diameters that subtend an angle equal to or greater than some specified value, viewed from a fixed point, is counted. The BA per hectare is then that number multiplied by a basal area factor based on the limiting size chosen. Wood volume is estimated using whatever formula is considered best for the type of forest under study. Since the forester using this system can "cruise" and measure a large number of plots and a large area of forest quite quickly, this system is faster and more efficient than measuring sample plots and suitable for covering large areas of forest. However, when the measurements are made at short intervals the BA estimates are unlikely to be accurate enough to provide good estimates of stand growth rates. A good article on conducting a timber inventory to give stocking and a measure of the basal area timber inventory is provided by Henning and Mercker (2009).

Despite the vast amounts of effort that have gone into estimating the amount of wood in natural forests, the results are often unsatisfactory and can lead to disputes, as in the case of the Tasmanian forestry industry. Modern techniques based on remote sensing are making progress toward solving this problem (discussed later).

Stand growth rates can, in principle, be estimated from measurements made at intervals on the same trees: growth is the volume at measurement 1 (V_1) subtracted from volume at measurement 2 (V_2), that is, $\Delta V = V_2 - V_1$. It is standard practice to establish permanent sample plots (PSPs); their locations are mapped and, in many cases, individual trees are identified so that the growth of those trees is documented by sequential measurements over time. In some parts of the world there are PSPs in natural forests in which the trees have been measured many times, usually at intervals of between five and ten years, over many years. The data from these measurements have provided an enormous amount of knowledge about forest growth. PSPs have also been established in plantations, where the measurement interval is usually shorter.

Not all the trees in a stand survive to maturity; there is mortality. Trees are counted when plots are measured at intervals to provide a record of changes in populations over time. Trees may die as a result of self-thinning—natural mortality in heavily stocked stands that occurs as a result of competition for resources, particularly light and water. Smaller trees are shaded out as stands develop and the larger trees dominate the canopy. Mortality may also be caused by disease or insect attack, storms or wildfires.

The large amounts of data produced by forest inventory measurements have given rise to a great deal of very sophisticated statistical analysis. A variety of nonlinear mathematical relationships has been used to describe the time course of volume growth and stem mortality. Statistical treatments allow evaluation of the errors and uncertainty associated with the final estimates of growth rate and wood yield and stem numbers. Some quite difficult problems may arise in determining growth rates because growth is calculated as the difference between sequential sets of "noisy" data. To illustrate: if the average stem diameter of a set of trees at time t_1 is 300 mm and a year later the average diameter is 350 mm, the apparent growth rate (Δd_B) is $d_B(2) - d_B(1) = 50$ mm yr^{-1}. But if the uncertainty in the values of d_B is, say, 3 mm, then the uncertainty in Δd_B is $\pm$ 6 mm, that is, 12 percent.

The productivity of an area of forest is commonly characterized by the *site index*. Site index, which reflects the fertility of the area as well as its climate, is estimated in many ways. The most common approach, particularly in even-aged single-species stands, is to measure the heights and diameters (at breast height) of the dominant or largest-crowned trees at a specified age. The twenty tallest trees at, say, age twenty, might be selected. The height is then converted to a yield class—a measure of wood volume production—using equations established from mensuration studies or, where data from long-term measurements are available, from look-up tables. *Yield tables* are often developed for a range of site indices and forest types. They essentially list the expected productivity, in terms of volumetric yield of forests on sites of specified quality (site index). Yield tables provide managers with a quick and convenient means of estimating standing volumes and average growth rates, or *mean annual increment* (MAI, m^3 ha^{-1} yr^{-1}) for fixed periods of five to ten years (periodic mean annual increment) or for an entire rotation.

There are several problems with conventional mensuration-based forest models. First, it's expensive and time consuming to set up sample plots and make enough measurements to provide useful statistical relationships. Sample plots, as the name implies, must represent the spatial variation across the forest(s) with which we are concerned, which means there has to be survey work in stands of different ages to identify differences in land units and site index. The costs are ongoing, since measurements must be repeated. Second, the relationships established are unique to the forests from which the data were obtained; it's not safe to extrapolate to sites with different soils, climates, and forest establishment or regeneration dates or management regimes. Separate models have to be developed for each location. Empirical models are also dependent on the weather conditions that pertained during the periods covered by the measurements analyzed; it's not safe to extrapolate into the future when weather conditions are likely to be different (recall chapter 4). However, despite these problems, mensuration-based statistical models have been the backbone of forest management for more than one hundred years, and vast amounts of useful data have been collected during their development. Where they have been archived in accessible form these data provide a valuable reference base, which can be used in the development of improved models.

The information about growth and yield that the forest manager needs in order to make economic and operational decisions is encapsulated in figure 6.1. The data in this figure were derived from a process-based model called 3-PG (the acronym stands for Physiological Processes Predicting Growth), which we developed and have described in the technical literature (Landsberg and Waring 1997; Landsberg and Sands 2010). It has been calibrated and extensively tested for the species concerned. Process-based models are discussed in the next section.

The figure is in three parts: figure 6.1a shows the volume of wood in a fertilized eucalyptus plantation in Australia over a nine-year rotation period together with the derivative of the curve in terms of time, that is, the change in volume per unit time. This gives the *current annual increment* (CAI) in m^3 ha^{-1} yr^{-1}. The yield at the end of the rotation is about 220 m^3 ha^{-1}. CAI peaks at about the halfway stage of the rotation, slowing from about 33 m^3 ha^{-1} yr^{-1} at year 5 to 22 m^3 ha^{-1} yr^{-1} at the end.

Figure 6.1b shows the curves for a totally different type of stand—Douglas fir in Oregon, with a stand age of 150 years, when the wood volume is almost 1,700 m^3 ha^{-1}. Peak CAI, about 23 m^3 ha^{-1} yr^{-1} is reached early in the life of the stand, after about 30 years. In figure 6.1c we have normalized the two sets of data from 6.1a and 6.1b (i.e., divided each value by the highest value of that set) and plotted them on the same graph to illustrate the similarities and differences in the growth patterns. The yield curves fall virtually on top of each other, but the CAI curves, as we noted, are displaced.

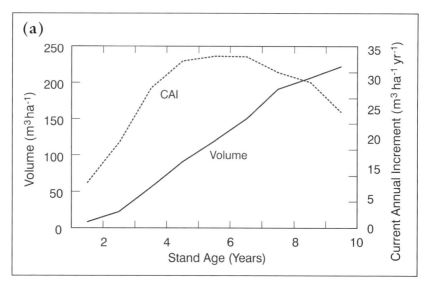

FIGURE 6.I. (a) Growth of a short-rotation (9 years) eucalypt plantation. Growth rate (CAI, m³ ha⁻¹ yr⁻¹) is high and remains high to the end of the rotation, but the stand accumulates far less total volume than the Douglas fir. (b) Growth of an unmanaged stand of Douglas fir in the Pacific Northwest, USA. The CAI peaks at about 23 m³ ha⁻¹ yr⁻¹ at about age 30, then declines steeply, but remains positive for the whole 150-year growth period, producing high final wood volume. (c) Relative performance of the two contrasting types. The curves are the ratios of CAI and total volume, plotted against normalized stand age, that is, (age/final age).

Process-Based Models

Process-based models (PBMs) are sometimes called mechanistic models. Unlike empirical models, they attempt to describe the behavior of the system in terms of the processes that drive it. The central problem in formulating a PBM is to capture the essential features of the most important processes. There are many such models in the literature, but very few are used for anything other than research.

The 3-PG model (Landsberg and Waring 1997) has been widely adopted as a research tool and is being tested as an operational or management tool in several countries. It is already being applied as a management tool in several South American countries. 3-PG is a relatively simple model that we produced with the explicit aim of bridging the gap between detailed models that try to capture all (or

(b)

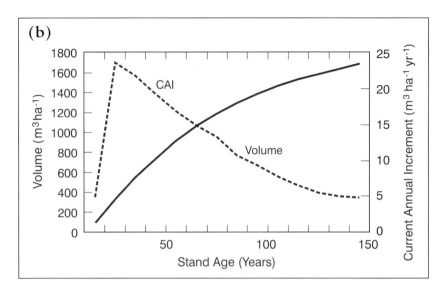

(c)

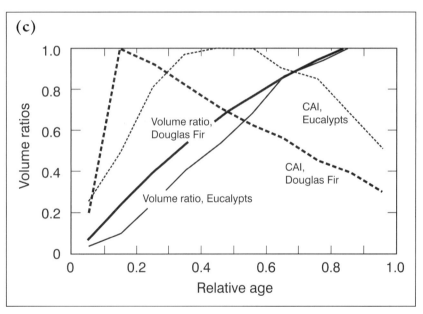

most of) the processes involved in the growth of trees and forests, and empirical growth and yield models. It's a hybrid model, consisting of a series of mechanistic submodels that describe the biophysical processes that govern the growth of forests (see chapter 3), and an empirical component that describes the geometry of trees and allows the calculation of stand volume and mass. The mechanistic components are consistent with those of most detailed models, so the following description provides an outline of the essential features of the family of PBMs. The principles underlying the model, and its structure, have also been described in some detail by Waring and Running (2007: see their chapter 3). Landsberg and Sands (2010) have described it in great detail and reviewed its performance in a number of countries, on a range of species in different applications.

Working on a monthly time step, 3-PG is driven by weather data and uses information about soil fertility and water-holding characteristics, while different tree species are characterized by the values of key parameters that control the physiological relationships in the model. Stand characteristics are provided as starting conditions; they can be anything from seedlings to mature stands. The amount of carbohydrate produced by photosynthesis is calculated from the amount of short-wave radiation absorbed by the forest canopy (dependent on LAI). The amount of biomass produced per unit of absorbed radiation, that is, the photosynthetic efficiency of the canopy, and the rate of transpiration, are affected by temperature, atmospheric CO_2 concentration, soil water content (obtained from water balance calculations), and the vapor pressure deficit of the air (acting on the stomata; see chapter 3). Carbohydrate is partitioned to foliage (some of which falls), stems and roots by a procedure constrained by robust and conservative *allometric* relationships. Soil water balance takes into account canopy leaf area and the effects of climatic and site factors on transpiration. The model includes a stem mortality function, and leaf litterfall. Stem mass and volume are updated monthly.

Stem mortality may also be induced in the model by thinning to different diameter classes. If insects or disease cause defoliation or kill trees at known intervals and intensities, the model has the capability to assess these effects on LAI and subsequent growth.

The widespread adoption of a PBM such as 3-PG as a practical management tool will take time, particularly where rotations are long and good historical mensuration data exist. Many foresters and managers are doubtful about using a model based on biophysical processes that seem to them somewhat esoteric. However, the prospects of increasing adoption seem favorable as the number of species on which 3-PG has been tested increases steadily, and it is seen to perform well as a predictive tool. It has the advantage that it can take account of the effects of variable weather and soil conditions and increasing CO_2 concentrations on forest growth and can be used to assess the consequences of management actions such as thinning. In countries interested in the establishment of nonnative trees in plantations, PBMs provide a means of matching species to environments where they are likely to do well. We do not claim that 3-PG is the final answer to the problems of predicting the growth and yield of forests, but it is being constantly improved and already provides a valuable and powerful tool to assess management options.

Remote Sensing Measurements

The principles underlying 3-PG—the conversion of intercepted radiation to biomass—can also be applied to data obtained from earth-observing satellites carrying instruments that now provide complete coverage of the earth's surface. There are instruments on satellites that monitor the radiation reflected from surfaces in the visible and near-infrared wavebands. The amount of foliage carried by vegetation in any pixel can be estimated from the reflectance ratios in different wavebands. The optics in satellite sensors can also indicate differences in the greenness of the leaves, similar to the way our eyes detect those differences. Conventional physiological research has shown that leaf greenness can be equated with different nitrogen content in the leaves, which is associated with changes in photosynthetic capacity. The water status of forests—that is, whether they are well watered or suffering from drought—is also indicated by the reflective properties of the leaves and the emissivity of long-wave thermal radiation from the soil surface. So the development of drought, which affects the uptake of carbon dioxide by vegetation, can be identified from sequential scans and factored into estimates of photosynthetic rates.

Estimates of the amount of foliage per unit ground area provide estimates of LAI, and we can use the fact that satellite measurements are repeated at regular intervals to follow long-term trends in LAI, noting whether they increase, decrease, or remain stable. Satellite measurements also provide data on solar radiation, and the repeated measurements allow us to keep track of changes in greenness and of indices of plant water status. Land-surface temperatures can be estimated from their emissions of long-wave radiation. Putting all this information into our models, combining it with ground-based weather measurements, and integrating over seasons, we can calculate how much carbon dioxide has been fixed by the forest and used for aboveground and belowground growth.

We can also estimate forest mass from an aircraft-mounted instrument called LIDAR (Light Detecting and Ranging), which projects laser beams directly downward. The beams bounce back to receivers with intensities that vary depending on the amount of foliage they strike, or on whether they hit gaps in the canopy. The differences between reflections from the top of the canopy and from gaps provide estimates of forest height as well as how foliage is distributed. Radar, projected both sideways and downward, can provide an accurate picture of the three-dimensional structure of forests. The return signals from radar of different frequencies reflect differences in the size of objects in their pathway. Small objects like leaves return higher-frequency waves. Larger objects, such as tree trunks, reflect lower frequencies. Combining all this with data from LIDAR, it becomes possible to estimate the aboveground biomass (by knowing the density of stemwood), and hence the carbon content of the forest stands.

These satellite-derived estimates of tree heights and biomass are being calibrated, and their accuracy tested, in research projects in various places around the world, where the results are being compared to conventional, ground-based measurements of tree volume and mass. The areas where measurements have been made on the ground are identified by what the remote sensing people call georeferencing—made much easier nowadays with *Global Positioning Systems* (GPS). Agreement between directly measured and satellite-derived estimates of forest mass is generally excellent. Progress in combining process-based models and remote sensing technology is leading us to the point where we can obtain good estimates of forest biomass and, based on sequential satellite measurements, which, except for LIDAR at present, are routinely available, forest growth for any specified area of land over known time intervals.

There is a further advantage in using satellite technology, and that is the opportunities it offers to monitor the condition of forests. The waveband composition of reflected radiation changes as foliage dries out, so sequential measurements can provide an index of moisture stress and its effects on the foliage of trees, and hence on growth. Similarly, damage to canopies by insects or disease shows up as differences in radiation reflectance.

It's unlikely that those who manage relatively small forests or units of forested land will be using satellite technology in the very near future, but this technology is already at the point where companies managing large areas, or those concerned with state, and certainly with national, forests, would do well to take advantage of its capabilities. One of the most important applications of remote sensing is as a mapping tool: different land (and water) surfaces result in differences in reflected or emitted radiation, so the land can be mapped as homogeneous cells (pixels). Pixel sizes, and hence the detail that can be obtained, depend on the instrument being used. The information that remote sensing can provide is of great value to policy development and decision making. Where remote sensing is used by forest managers, experts in the analysis and interpretation of the data will still be required, but it seems likely that software packages that do most of the number crunching will be available in the not too distant future. Waring and Running (2007) have provided a quite comprehensive and detailed treatment of the use of remote sensing in ecosystem analysis at different scales.

Fire and Its Management

Forest fires are natural ecological phenomena. They are often started by humans—either by accident or intentionally—but they are also started by lightning. The challenge facing the forest manager is to control fire and use it as a tool that contributes to sustaining the quality and ecological properties of the forests.

Until the second half of the twentieth century, government policies concerned with fire in forests—called wildfires in North America, bushfires in Australia—were aimed at immediate and complete suppression. The general view was that fires damaged and possibly destroyed forests, and they are certainly dangerous to human dwellings and often to human life in fire-prone areas. Until about 1970 the US Forest Service had a policy of complete fire suppression; the Smokey Bear advertisement ("Only you can prevent forest fire") has been called one of the most successful advertising campaigns in history. The approach in Australia is, in general, suppression—or at least containment—where possible. All the Australian states maintain fire services, mostly based on volunteers, but every summer brings uncontrollable and sometimes disastrous bushfires.

The fire suppression policies followed in North America and Australia were (and are) expensive in terms of manpower and resources such as trucks, aircraft, and communications systems. They have also had unforeseen consequences. It's almost impossible to prevent wildfires in perpetuity, and prevention over periods of years results in dense stands and the accumulation of large amounts of dead biomass: fallen trees, branches, twigs and, in drought periods, dry leaves that don't quickly rot. This material provides fuel, so that when a fire occurs in conditions in which it cannot be extinguished, its intensity is likely to be high and much more damaging than it would have been if the forest had been burned more frequently. The high-intensity fires that occur in extreme fire weather—hot, dry, and windy conditions—tend to be crown fires: the oils in the foliage of eucalypts volatilize so combustion is spectacular, as it is in the canopies of conifer forests, such as those of North America and the boreal regions. These fires are likely to kill many of the trees, even in the case of species that are adapted to fire. Tree species have a range of responses to fire; for example, many of the eucalypts survive moderate- to high-intensity fires because of thick bark and *epicormic* shoots that emerge from under the bark. The thick bark of ponderosa pine allows it to survive moderate fires, whereas lodgepole pine is killed even by low-intensity fires but regenerates quickly from seeds released from the heated cones. The long-term effects of fire suppression policies are being exacerbated by the effects of climate change (see chapter 4).

From about the 1960s, evidence began to accumulate from ecological and historical studies that some forests need to be burned periodically to maintain the integrity of stands and ensure regeneration. The giant sequoia stands of California, for example, need bare mineral soil and a favorable light environment for successful seedling establishment and growth, and many eucalypt species regenerate best on ash beds. The Aboriginal people of Australia used fire as a management tool to produce fragmented landscapes that encouraged the growth of varied plant species and the movement of animals into particular areas where they could be easily hunted. Evidence from charcoal and fire scars on ancient tree stumps, as well as the experience of the early European settlers, indicates that the landscapes recognized as distinctively Australian were formed by fire over many human generations.

It's now generally accepted that, in many forest types, periodic low-intensity, controlled, fuel-reduction burns are an important part of the management of natural forests, not only as a means of reducing the danger from high-intensity fires but also for ecological reasons. The objective is to reduce undergrowth and vegetation that would compete with regenerating trees, reduce fuel loads and logging slash, and improve wildlife habitat. These fires are deliberately lit at times when the forest is dry enough to burn, but not so dry that the fire is likely to escape. Wind conditions must be moderate, and the area to be burned is normally bounded by fire trails or natural breaks that allow it to be contained. The frequency with which forests are burned should be related to forest type and condition, but the different permutations arising from alternative options and possibilities, influenced by ecological considerations, requirements for resources, climate, and politics, are almost endless. Fuel reduction burns are not usually used for species like lodgepole pine or fir species that typically do not experience low-intensity fires.

Fire is commonly used as a management tool to get rid of unwanted slash (debris) after a stand is harvested. It may result in the loss of plant nutrients from the soil, particularly nitrogen and sulfur, which are volatilized during combustion of organic matter. However, nutrient losses are likely to be offset by increased mineralization, making nutrients such as phosphorus and potassium more easily available to plants. Fire is quite widely used to "clean up" former plantation sites prior to reestablishment.

It's not always possible to carry out controlled burns at the optimum time, and when trees weakened by drought are killed by disease or insect attack, as is happening in the western states of the United States and Canada, it becomes difficult to carry out controlled low-intensity burns. The forests then become increasingly susceptible to severe fires, as evidenced by the increasing number of fires in those areas in recent years.

An alternative approach to controlled burning is the "let burn" option. This means that when fires occur as a result of lightning, or even accidental (or sometimes malicious) human action, no attempt will be made to put them out. This option is often applied in national parks, where the costs of trying to suppress fires are high, and the actions taken to do so are frequently ineffective. In most cases the forests will recover. However, this is not a policy readily accepted by the public, particularly those who, as we noted in chapter 5, have built houses or weekender cabins in forested areas. These people are also generally unsympathetic to the idea of controlled burns in their vicinity. Investigations in Australia have indicated that clearing the vegetation surrounding these buildings, making it easier to fight fires around them, is the best safeguard against their destruction by fire. But even this is not popular because it's seen to detract from the sylvan attractions of the environment. Consequently, managing fire and its impacts in the "wildland-urban interface" is politically and sociologically difficult in both America and Australia. Canada follows fire suppression and control policies in its more accessible forest areas, but in the more remote boreal regions, where access is difficult and human habitations rare, let burn policies are likely to be followed.

Fire control policies, and the management of fire in forests, are never going to be easy. The situation is constantly changing as forests grow and human habitations push into forested lands. Climate change and the increasing frequency of drought and high summer temperatures in many fire-prone regions are complicating factors. The ecological requirements for sustainable management may not be consistent with social and political attitudes. In the United States there are tensions between federal authorities—mainly the Forest Service, on whom the burden of fire control falls—and state agencies, revolving around budgets and fire-fighting resources. Those who manage forested public lands can only develop policies and strategies best suited to local environments, consistent with the requirements for sustainable management, and try to ensure that the fire management practices they follow adhere as closely as possible to those strategies. Much of the time, strategy and practice will diverge significantly.

Summary

Managing forests involves making decisions that are very often governed, or at least strongly influenced, by economic considerations. We have divided economics into macroeconomics, concerned with the economies of countries, and microeconomics, which, in relation to forestry, are concerned with the economies of companies and individual properties. The principles of classical economics don't work very well for forest management because of the long response time of production processes relative to market demands, as well as for the usual reasons of imperfect knowledge and the nonrational behavior of markets. Forests constitute valuable natural capital, which is generally treated as an uncosted externality in national accounting systems and calculations of the gross domestic product. We consider, very briefly, forest management policies in the United States, Australia, Canada, Sweden, and Russia, and in the tropics.

At the microeconomic level, we analyze the costs incurred by logging companies working on public land. To illustrate the situation in relation to forestry on private land, where the cost of the land adds to the operating costs, we use an elementary model of a large plantation operation. Discounting has to be used to estimate the likely returns at harvest from forest operations. To do this it is necessary to estimate the yields likely to be obtained from forest stands, both natural and plantation. This is normally done using empirical growth and yield models, but we argue that process-based models must gradually supplant these. Process-based models can be used in conjunction with the powerful techniques of remote sensing.

The last section of the chapter deals briefly with fire, its impacts and uses in forest management. The questions of fuel reduction, burning, and control of wildfires are discussed.

Chapter 7

The Future for Forests

Forests have been important to humans throughout their history; they are important now and will be important in the future. But forests of all types are under heavy pressures, brought about by changing climatic conditions and the impacts of human use and abuse. If we are to successfully mitigate and adapt to the pressures resulting from climate change, and stem the steady reduction of forest area around the world, we need science-driven political action. In view of the prevailing short-term and relentlessly materialistic world mindset,[1] this is going to be difficult to achieve. To make progress, we have to work on increasing, in politicians, planners, and decision makers, awareness of the fact that we are all both consumers and beneficiaries of the natural resources provided by the earth. Forests are not the least among these. We all share the consequences of decisions that affect those resources, either adversely or to our benefit. We also need to raise awareness that the global economy is inextricably linked to the ecological benefits provided by the earth's ecosystems, so changes to those ecosystems have far-reaching implications.

Political pressure, at least in democracies, comes from the people, and even in countries that are not democracies, where the people's interests are clearly compromised by the actions of government, there will be pressure for change. But where the factors affecting interests are subtle, and not widely perceived, so that most people are not aware of them, the pressures for change may not be strong enough to bring it about. This is likely to be the case in relation to forests, particularly among those who live in cities. Great numbers of people have never experienced the aesthetic benefits of forests, and they are not generally aware of the broad range of benefits that humans derive from forests. They know little about atmospheric CO_2 concentrations and the role of forests in helping check their rise (see chapter 4), and they barely register changes in the climate, which are masked by normal daily and seasonal fluctuations. The subtle but important differences associated with the loss of forest cover do not impinge on them. To people in the towns and cities of the developed world clean water comes from taps. The value and importance of forests in the watersheds that supply the reservoirs is not a matter of concern to them. Similarly, paper in all its forms is something you buy. For most of those who work with wood and wood products, their source is the hardware store or lumber yard: where the wood comes from is not their concern, although it will become a concern if prices rise too much.

But, despite this widespread ignorance and unawareness, environmental issues are becoming increasingly important in political, economic, and social debates. We said, at the beginning of this book, that one of our objectives was to provide science-based knowledge and information as a contribution to political and commercial actions and decisions about the use and management of forests. In trying to achieve this we have identified and discussed a range of problems and issues.

We saw in chapter 2 that forests vary enormously in their biological and ecological characteristics, depending on the climatic zones in which they are located, and on the soils and topography in those regions. The pressures to which they are subjected by humans range from outright destruction, sometimes intentional, sometimes inadvertent, to degradation and loss of the ecosystem values and services that healthy forests provide. Intentional destruction includes clearing for agriculture or for enterprises such as the oil palm plantations that are proliferating in Indonesia and Malaysia (chapter 5). Inadvertent destruction may be caused by management practices that make forests vulnerable to high-intensity fires or to insect or disease outbreaks, such as those that are increasing in Canada and Alaska as a result of human-induced climate change.

We covered in some detail the interactions between weather and forests (chapter 3), climate change and its effects and implications (chapter 4), and the ecosystem services we obtain from forests (chapter 5). In chapter 6 we dealt with forest management within the framework of the economic principles that influence or indeed determine the decisions that managers must make.

In this chapter, we consider the future of forests, taking account of the critical role and influence of climate change. We understand the causes of climate change, and could—at least conceptually—slow or halt it. This is not likely to happen, but forestry can make a significant contribution by reducing (preferably eliminating) biomass burning. Forest managers must also recognize that it is almost certain that the world's climate will continue to change too fast for native plants and their supporting fauna to adapt.

We discuss the reasons for conserving the world's natural forests, and how those forests should be managed so that they are not destroyed or degraded. And we consider the management of plantations, which are already important as the primary sources of forest products with direct economic value, and must become more so.

Conserve the Forests

We begin with the premise that, wherever possible, natural forests should not be cleared but should be managed sustainably. This is a particular challenge in the face of pressures on land from growing human populations. Climatic change and increases in atmospheric and water-transported pollutants, including nutrients, bring further pressures. Over time, there are likely to be changes in the composition of forests. The transition to new, sustainable ecological communities through maintenance of wildlife corridors across the landscape must become a component of sustainability.

The fundamental obstacle to forest conservation is, in many cases, the tendency in both the business and political worlds to make decisions and take action on the basis of short-term economic considerations, without taking account of the ecosystem services provided by forests (see chapters 5 and 6). There is profound and widespread ignorance about ecosystem services and the natural capital provided by forests, to the extent that (to our knowledge) it is not included in the national accounts or the estimates of the gross domestic product of any country. This distorts those accounts and estimates. It also means that, in many countries, natural forests are rarely managed with a view to preserving their capital value.

In terms of conventional economics, the returns likely to be realized by converting forests to agricultural land or by harvesting standing timber using destructive wholesale logging practices generally easily outweigh ecological and environmental considerations. This obstacle to sustainable forestry will only be overcome by a mixture of public pressures and regulation, but its removal would be accelerated by accounting procedures that take cognizance of the value of ecosystem services and natural capital.

Recent information published by the (United States) National Academy of Science[2] revealed the slightly surprising fact that, in the period 2000–2005, the United States lost a higher proportion (6 percent) of its forest cover than any other of the seven nations with the largest forest areas in the world (Russia, Brazil, Indonesia, Canada, China, Democratic Republic of the Congo, and the United States). Canada was second, with 5.2 percent, followed by Brazil (3.6 percent), Indonesia (3.3 percent), Russia (2.8 percent), and China (2.3 percent). Brazil (165,000 km²) and Canada (160,000 km²) lost the greatest areas of forest in absolute terms. By forest type the greatest losses were in the humid tropical forests, followed by boreal, dry tropical, and temperate, in that order.

Whatever the precise figures, these losses are enormous, have serious consequences, and are, in the long run, unsustainable. They encapsulate huge destruction of wildlife and the loss of unique floras, along with critical ecological services. There were almost certainly species of fauna and flora in the areas where forests were lost that were adapted to specialized ecological niches but became extinct as a result of habitat destruction. The implications in terms of loss of carbon storage, changes in hydrological regimes, and direct contributions to global warming through alterations of the energy balance are also significant. In the United States, Canada, and Australia, progress in stopping the destruction of forests may be made by political campaigns and action to educate the public, most of whom are unaware that the rates of forest loss have serious implications for the well-being of future human generations. Australia is a minor player in terms of forested areas, but large areas of woodland continue to be cleared there. Recent increases in clearance rates in Queensland are justified by the argument that the state is in drought and farmers must be allowed to clear land to increase the areas of grazing available to their cattle. So short-term economics again trump environmental considerations. There are also continued logging, involving clearcutting (clear-felling), in old-growth forests in Tasmania and Victoria, and efforts to expand logging in the national forests of the US Pacific Northwest. The rationale, in all cases, is the need to support the forest industries and create jobs in forest-dependent communities.

Forest losses in the humid tropics are particularly serious, for three important reasons: (1) The destruction of tropical forests results in massive losses in biodiversity—many conservation biologists believe that an extinction crisis is looming in these forests. (2) The carbon emissions resulting from burning the slash left over after harvesting tropical forests, or removing them to make way for oil palms or agriculture, constitute about 15 percent of global emissions—more than the annual emissions of either

China or the United States. (3) About 1.5 billion of the poorest people in the world, in Southeast Asia, Brazil, and the Congo—more than one-fifth of the world's current population—rely on forests for their livelihood. Most of these people have no title to their land, so when the governments of those countries sell the land for agriculture or allow harvesting by foreign companies with no concern for the health of the land or the well-being of the local people, then those people may suffer grievously.

Forest destruction in the humid tropics (see chapter 5) is partly a result of loose or nonexistent regulations and control. Licenses are issued to foreign companies, many operating from Malaysia, Korea, and China, to log large tracts of forest (logging concessions). They are generally unsupervised. The extraction techniques used are likely to be highly destructive and leave the land degraded and susceptible to erosion from the roads pushed through the forest without concern for topography or soils. There is also the problem of illegal logging. Bill Laurance[3] says that, according to a 2011 study by the World Bank, two-thirds of the world's top tropical timber-producing nations are losing at least half of their timber to illegal loggers. In some developing countries the figure approaches 90 percent. The loggers get around regulations and laws by falsifying logging permits and using bribery to obtain illegal logging permits, logging outside of timber concessions, hacking government websites to forge transportation permits, and laundering illegal timber by mixing it in with legal timber supplies. Illegal logging returns no taxes to the countries that suffer from it, and very little in terms of wages to the locals, who are often brutally exploited. It can only be countered by increasing international support for forest and wood certification schemes that aim to ensure that wood products are being grown, harvested, processed, and sold sustainably. Like so many schemes involving international trade, reaching agreement on principles, standards, and enforcement is fraught with difficulty, but it is important that the efforts to do so are supported.

A scheme called REDD— *Reducing Emissions from Deforestation and Forest Degradation*—aimed at reducing emissions from forest conversion and conserving carbon stocks, emerged as one of the few points of agreement between nations at the 2009 Copenhagen Climate Change Conference. This scheme relies on the world's rich nations to contribute US$10–$15 billion annually to provide economic incentives to stop deforestation, including by stopping illegal logging, and by promoting other means of economic development in forest-dependent communities. Unfortunately, it seems that, as with other such schemes, the willingness of the rich countries to contribute funds has not matched their rhetoric when discussing the matter. But, sometime in the not too distant future, such schemes must be made to work.

Management for the Future: Natural Forests

A paper by Aber et al. (2000) provides a valuable guide and framework for discussion of the management of natural forests in countries where laws and regulations are respected, and where environmental awareness is strong enough to encourage—if not always enforce—obedience to those laws and regulations. The paper is part of a series on "Issues in Ecology," produced by the Ecological Society of America (ESA). It's concerned with the management of native forest ecosystems and outlines the key ecological considerations that should underlie sound forest management. The authors subscribe to the principle of sustainable management, which they define as the management policies and practices that do (should) not devalue the resources for future generations. They place strong emphasis on the maintenance of soil quality and forest health, noting particularly the phenomenon known as "nitrogen saturation," which occurs when rates of input of nitrogen-rich pollutants into soils exceed the rate at which plants and micro-organisms can use or store it. When that happens the excess is lost to streams, groundwater, and the atmosphere, with deleterious effects. Aber et al. also note that plant roots are important for slope stability; forestry practices that decrease root density destabilize slopes and contribute to slope failures—witness the sometimes disastrous landslips that occur during heavy rain in areas where the slopes have been deforested.

Planning at the landscape level is needed to address ecological concerns such as biodiversity and water supply and quality. To maintain biodiversity, forest fragmentation caused by clearcuts and roads should be avoided, and vulnerable areas such as old-growth stands and riparian zones should not be harvested. There is no scientific basis for the assertion that silvicultural practices, such as clearcutting "patchwork" areas within a forest, can duplicate or substitute for the ecological effects of natural disturbances, or create forests that are ecologically equivalent to old-growth forests, although techniques such as retaining (seed) trees and large woody debris in harvested areas can come close to mimicking natural processes (Aber et al. 2000).

The ESA panel was concerned only with American forests, and their recommendations were aimed at the management of those forests, but they provided a set of guiding principles that can be applied to the management of native forests elsewhere. However, we need to consider the economic as well as the ecological aspects of forest mangement in the future. There are some useful papers dealing with this topic in the journal *Forests* (vol. 2, 2011), where Nordin et al. point out that the management of forests in the future must be able to deal with tradeoffs between different ecosystem services when these come into conflict with one another. For example, there may be conflict between demands for timber for pulpwood, energy, or sawn timber from forested areas that are also important for recreation and clean water supplies. In such cases, it's not likely that an optimum solution, where all the services required are maximized, will be possible. Management prescriptions will have to be decided by a multidisciplinary approach based on discussions between people of different disciplines and stakeholders with a range of interests. Nothing will be achieved by dogmatic insistence on the approach suggested by a particular discipline.

There are two well-developed techniques, now being widely used and tested by research scientists but not yet by managers and decision makers, that must become important tools in forest mangement in the near future. These are process-based models and satellite-based remote sensing. They are both powerful in their own right, but even more so when used together. It's becoming clear that predictions made using conventional, mensuration-based growth and yield models (chapter 6) are becoming, and will increasingly become, less and less reliable. These models are developed using measurements made over long periods. They rest on the assumption that climate is stable, that is, that the climate in the future will be similar to that of the past, so that extrapolation into the future is justified. They also assume that attacks by insects and disease organisms in the future will follow (with allowance for natural variation) the patterns of the past. Both assumptions are now untenable. Forests are entering states of continuous transition; climate is changing and with it the growth patterns of trees and forests. Process-based models that incorporate the mechanisms that determine reponses to weather and climate (see chapter 3) allow those responses to be predicted. It's also possible to include in these models the damage and response mechanisms associated with insect attacks and tree diseases. Good process-based models already exist, and many of the principles underlying them are well understood. There must be increased focus on developing them in user-friendly forms, so that managers and decision makers are comfortable using them. They have the additional advantage that they can be used as heuristic tools, to explore the consequences of various "what if?" questions and likely scenarios.

We noted, in chapter 6, that measurements made by remote sensing from satellites or aircraft (satellites are more useful because they provide continuous data[4]) can be used to estimate forest mass, health, and growth potential: What is the leaf area index? How green is the canopy? Are there temporal changes? Combining these measurements with ground-based weather measurements and using appropriate process-based models, allows detailed, short-term calculations of forest growth that can be checked using remote sensing. Furthermore, this can be done for large—indeed vast—areas. Remote sensing data are routinely stored in geographical information systems (GIS), where the measurements may be combined to provide maps consisting of cells (pixels), each of which contains several layers of information—soil, forest type, stand condition and biomass, climatic data, and so on. It's essential that the use of these tools becomes routine, so that growth and yield estimates become available for all the world's forests and disturbances can be continually monitored. Remote sensing technology provides the means to map changes in land use; it is a powerful tool for monitoring forest destruction and degradation. Landsberg and Gower (1997) provided a quite detailed description of the use of remote sensing and GIS in forest management and modeling, and Waring and Running (2007) provided a definitive and complete treatment of the analysis of forest ecosystems at multiple scales.

Management for the Future: Plantations

Turning to plantations, we noted in chapter 2 that, while they represent less than 3 percent of the world's forest cover, they produce more than 25 percent of our wood products. We can expect this proportion to increase and, in the interests of conservation, it must do so, although not through conversion of more native forests. In chapter 6 we discussed the management of a plantation estate in economic terms, but said little about the sivicultural and ecological aspects.

Plantations are crops and should be managed for high and sustainable production, which means a strong focus on good soil management. Erosion must be prevented, particularly when trees are young, and soil organic matter and fertility must be maintained. The losses of fertile soil that have taken place as a result of ill-advised and exploitive agricultural practices in many parts of the world, such as the Midwest of the United States, the wheat-growing areas of Australia, and large parts of India and China, are catastrophic. They cannot be reversed and, in this age of rapidly increasing human populations, they have serious implications for food production. The mistakes made in agriculture must not be repeated in plantation forestry.

We noted in chapter 2 some of the countries where plantation forestry is already important and, in some cases, becoming more so. We also mentioned there some of the disadvantages that might be associated with plantation culture: limits to biodiversity, possible alterations to the local hydrology, possible runoff of herbicides and insecticides into local water supplies, and unintended damage to native flora and fauna. Plantations are often planted on farmland, where the farmer has decided that they are likely to be a more profitable form of land use than some other form of cropping. In parts of the southern United States, such as Georgia and the Carolinas, loblolly pine plantations have replaced the cotton that was once the dominant crop. Since many of the old cotton lands were degraded, this is an excellent development from all points of view, especially in regard to wildlife. We expect to see increasing plantation establishment in many countries—it's already happening in China, Vietnam, Thailand, and, no doubt, other Southeast Asian countries.

There are a few general principles that need to be followed to ensure that plantations are sustainable into the future. Healthy and productive plantations require high LAI consisting of healthy foliage with adequate nutrition. To achieve this, attention must be paid to soil fertility and structure. Practices such as the use of leguminous cover crops between rotations add nitrogen to the soil and, plowed in or allowed to decompose, improve organic matter content and soil structure. This approach is widely used on the sandy soils of South Australia, where the early radiata pine plantations depleted the rather poor, sandy soils to the point that the growth of the second rotation trees was very poor. Two practices helped solve the problem. The first was to discontinue burning the slash (bark, foliage, and branch residues) left after harvest; instead it was macerated and partly incorporated into the soil using heavy machinery. The second was planting lupins (a legume, which fixes nitrogen) into the decomposing slash. The sandy soils tend to be poor in nitrogen, which is readily lost by leaching. Soil nitrogen is restored by incorporating the lupin crop into the soil before trees are reestablished. Such practices help maintain soil organic matter, fertility, and structure and, with conventional fertilization, ensure that growth and yield of the plantations can be maintained indefinitely. In other areas, reestablishment practices should, obviously, be tailored to soil type and the selected silvicultural regime.

Practices that help ensure that plantations do not degrade the landscape include avoiding riparian zones—stream and river banks—by leaving strips of native vegetation undisturbed. These strips should be wide enough to prevent streambank erosion and provide refuge areas for birds and other vertebrates, and for the invertebrates that play such important roles in protecting and regenerating native forests. Plantations should not be established in small catchments. Except for the last sentence, these remarks also apply to natural forests intensively managed for wood production, as is the case throughout the Nordic countries.

Large areas of single-species natural forest or plantations are not resilient systems. Recapping what we said about resilience in chapter 1, the threshold, or tipping point, is the point at which a system may move into a new state from which it will not return. The transition from one stable state to another may come about as a result of ecological or socioeconomic factors, so when we evaluate the resilience of plantations we have to consider their stability as ecological systems and as viable economic operations. Ecological stability may be affected by human-induced climate change (chapter 4), resulting in stress that may cause forests to flip into a new state from which they may not rebound. For example, they may become highly susceptible to disease and insect attack, as has happened with the lodgepole pine forests subjected to bark beetle attack in the western United States and British Columbia. Fire in these dead and dying forests is likely to bring about long-lasting changes in their structure and ecology. Similarly, Dothistroma, a fungal disease that affects a wide range of pine species was, until recently, mainly a problem in the Southern Hemisphere, but from about 1990 it has become increasingly serious in Britain and across the United States and Canada. It can be controlled by regular spraying with copper-based fungicide, but this is an expensive procedure, and it may not be possible to maintain it across all affected areas. Some forests may collapse as a result.

A quite dramatic example of the collapse of a plantation enterprise that was not economically viable occurred in Western Australia in recent years. Heavy investment in large areas of blue gum (*Eucalyptus globulus*) plantations was encouraged by federal tax incentives through so-called management investment schemes. The levels of production achieved in the first rotation of these plantations were, in many cases, much lower than the commercial hyperbole behind their promotion suggested. There was little excuse for this as well-calibrated models such as 3-PG (chapter 6) and ProMod (Battaglia and Sands 1997) existed and could have been used to evaluate the claims made. In some cases, where such models were used, the predictions of poor growth on marginal lands in drought-prone regions were ignored by those promoting the schemes. The situation became worse when plantations moved into the second rotation. In many areas the first rotation had grown largely on water stored in the deep soils. But the rate of water use by the plantations generally exceeded the rate of resupply by rainfall, a situation exacerbated by the fact that Western Australia is being seriously affected by climate change and is experiencing severe and prolonged drought (see chapter 4). The trees, therefore, suffered water stress serious enough to check growth almost completely and, in many cases, to cause death. Consequently, large areas planted to blue gums are being pulled out and will not be replanted. There is a range of reasons for this debacle, besides the fact that no account was taken of water balances and climate change—and in some cases, expert advice. They include poor land management, in terms of organic matter and fertility maintenance. The long-term effects on the soils may affect subsequent land use for many years.

The solution to the problem of maintaining resilience and long-term sustainability of single-species forests and plantations must lie in diversification: widening the range of genetic material and planting mixed forests with a variety of species. This will not find favor with the managers of large estates, since the costs of replacing their main species, even over a number of rotations, will be high. The costs of moving away from monocultures will also be high in terms of changes to production practices. This solution may drive small growers out of business. Nevertheless, the use of a much wider range of genetic material, in place of single-species monocultures, will become essential in most plantation-growing areas in the near future. A range of genotypes, with varying susceptibility to changing climates, will increase the resilience of tree-growing operations, making it less likely that adverse conditions will lead to tipping points and the failure of an enterprise. It will also be important to ensure that soil fertility and quality are maintained.

Concluding Remarks

We have, throughout this book, discussed forests, their importance, and their management in rather general terms: we have outlined principles and made valid generalizations, but we have not provided many detailed examples and case studies to illustrate the points we have made. In our treatment of forest types, we noted that differences in topography, disturbance (including harvesting), and fire lead to considerable variation within any broad category. People concerned with the use and management of forests in particular areas will need to understand and take account of these differences.

Similarly, we discussed the effects of climate on trees and forests, providing the background needed to understand the effects and interactions. Climate change was discussed as the global phenomenon that it is, and some of its implications in relation to forests were considered, but these implications will almost certainly be wider than we have indicated. The current and probable future effects of climate change will need to be factored into thinking about the future of forests and how they are managed.

We have recognized explicitly that humans will continue to use forests—there is no prospect that they can be locked up. Nor should they be. But we have to ensure that, as human populations and their demands grow, the forests survive with the properties and characteristics that make them of such great value to us, both for their obvious material products—like wood, high-quality water, and carbon sequestration and storage—and for the aesthetic benefits they provide. To this end we need to ensure that externalities, usually ignored in economic assessments of forest values, must be included in evaluating the consequences of any form of forest use. And, as we have said earlier in this chapter, we must ensure that plantations and single-species natural forests are managed sustainably by protecting the land and its biota.

We humans can no longer afford to treat the world's forests as expendable.

Chapter 1

1. Hugh Thomson. "The Sherwood Syndrome." http://www.aeonmagazine.com/nature-and-cosmos/ hugh-thomson-britain-forests-myth.

Chapter 2

1. The term *clearcutting* is used in the United States to describe the removal of all the trees from a site. The term used in Australia is *clear-felling*.

2. Ruth Young. 2010. "Nothofagus: Relics from Gondwana." TalkingNature.com. http://www.talkingnature.com/2010/02/ biodiversity/nothofagus.

3. It's interesting to note that the development of localized variation is the mechanism generally accepted as explaining the evolution of intraspecies differences in the Galapagos Islands, famous for their crucial role in leading Charles Darwin to the theory of evolution and origin of species. It seems there has been enough localization in the Amazon basin for the same mechanism to operate there.

4. http://www.borealforest.org/world/world-sweden.htm. Ed Pepke. "Global Wood Markets: Consumption, Production, and Trade." http://www.fao.org/forestry/12711-e94fe2a7 dae258fbb8bc48e5cc09b0d8.pdf. J. K. Boyle, K. Winjum, K. Kavanagh, and E. C. Jensen, eds. 1999. *Planted Forests: Contribution to the Quest for Sustainable Species.* Kluwer Academic.

Chapter 3

1. Technically water moves across water potential gradients. Potential is the energy per unit volume and is a measure of the capacity of the water to do work, compared with the work capacity of pure, free water.

2. The prefix *iso* implies equality, so isohydric species maintain their state of hydration at (approximately) some equilibrium condition.

3. The term *watershed* is sometimes used for catchments. Strictly speaking, a watershed is an area or ridge of land separating water flowing to different rivers, basins, or streams. A catchment is the area of land where water is collected and from which it flows.

Chapter 4

1. Walter (2010) provides a lucid explanation of Arrhenius's famous finding. His greenhouse law for CO_2 is $\Delta F = \alpha \ln(C/C_0)$, where C is the CO_2 concentration in parts per million by volume, C_0 denotes a baseline concentration of CO_2 and ΔF is the radiative forcing, measured in W m^{-2}. The letter α is a constant, which has been assigned the value 5.35.

2. http://environment.nationalgeographic.com/environment/habitats/last-of-amazonmagazine).

3. We generally use the word *significantly* in a statistical sense. In that sense it means that the probability of whatever event we are discussing is greater or less than some specified probability level, usually 0.95 or even 0.99. We will not, in most cases, provide actual probabilities: the use of the word implies that scientific statistical studies have been done and have led to results that indicate the differences under consideration cannot be attributed to chance. They are real.

4. Kate Galbraith, "Getting Serious about a Texas-size Drought," *New York Times Sunday Review*, April 9, 2013.

Chapter 5

1. Sustainability means that the capacity of the forest to provide goods and services of all types—not just in terms of wood production—should not be permanently diminished over time by any of the operations carried out in the forest (see chapter 6).

2. http://www.sciencedaily.com/releases/2008/07/080714162600.htm.

3. http://www.fao.org/docrep/011/i0410e/i0410e04.pdf.

4. http://environment.nationalgeographic.com.au/environment/habitats/rainforest-profile/.

5. The measure of wood density (mass dry material per unit volume) conforms to the conventional definition used in physics.

6. FAO Advisory Paper. 2007. "Global Wood and Wood Products Flow." http://www.unece.org/fileadmin/DAM/timber/mis/presentations/PepkeGlobalWoodMkts050510.pdf.

7. Ibid.

8. http://www.greenpeace.org/australia/en/what-we-do/forests/Forest-destruction/.

9. http://www.saynotopalmoil.com/.

10. "Final frontiers: Rainforests." The Conversation. https://theconversation.com/final-frontiers-rainforests-12922.

Chapter 6

1. Limited competition, as when the markets are dominated by a few large "players."

2. http://www.parl.gc.ca/Content/LOP/researchpublications/prb0513-e.htm#opposingtxt.

3. For example, a couple suing each other for divorce probably reflect nothing but unhappiness, but because they generate work for lawyers, clerks, and perhaps counseling services, their unhappiness contributes to the sum of economic activity and so increases GDP (although it presumably makes the lawyers better off!). Similarly, GDP is increased by a motor accident

that may result in injury or death, almost certainly causing unhappiness, wasting resources, and reducing social well-being. But it will provide work for repairers, police, perhaps hospitals, and possibly funeral directors.

4. http://cfs.nrcan.gc.ca/pages/132.

5. http://www.ncbi.nlm.nih.gov/pmc/articles/PMC3593033/.

6. In the United States, commercial wood volumes are expressed in board feet—the volume of a one-foot-long board, one inch thick, and one foot in width. However, this is not a recognized scientific unit and is not used internationally. We will stay with the metric system and use cubic meters and hectares ($1 \text{ m}^3 = 35.3 \text{ ft}^3$; $1 \text{ ha} = 2.47$ acres; $1 \text{ m}^3 \text{ ha}^{-1} = 14.3 \text{ ft}^3$ acre).

Chapter 7

1. Historically, of course, this is nothing new. For hundreds of years humans have been driven by materialism and the urge to own things. The difference now is that the avalanche of seductive modern goods, coupled with the ease with which transactions are made, and the efficiency of modern communications, have made the attitude almost universally pervasive. People everywhere expect that their standard of living will increase continuously.

2. www.pnas.org/cgi/doi/10.1073/pnas.0912668107.

3. http://theconversation.com/organised-crime-illegal-timber-and-australias-role-in-deforestation-10048.

4. However, we note that, at present, no LIDAR instruments are in orbit and, even when they are, it will be difficult to distinguish between conifer species.

Throughout the text, words that may not be familiar to all readers are italicized when first used. We have, in most places, defined them where they occur, but for convenience they are also defined here.

albedo. The proportion of short-wave energy from the sun that is reflected by the underlying surface.

allometry. The mathematical relations between components of a tree, specifically those between stem diameter at breast height and the biomass of stems, branches, and foliage.

angiosperms. Class of flowering plants with seeds positioned inside some receptacle, often a fruit. Leaves, even if evergreen, are generally replaced every year.

anisohydric. Trees can be grouped into *isohydric* and *anisohydric* types. Anisohydric trees do not exert tight control over the water content of their leaves and other tissues; stomata remain open, at least initially, even if transpiration rates are high and the soil in the root zone is becoming dry. This may lead to cavitation: air bubbles in the water-conducting vessels.

base flow. The portion of stream flow that comes from deep subsurface storage and delayed shallow subsurface drainage. Also called *sustained* or *fair-weather flow.*

bio-char. The term used to describe charcoal when it is used as a soil amendment to improve soil carbon content and water-holding capacity. Fertile soils are created in tropical areas by incorporating incompletely combusted charcoal along with massive amounts of green vegetation.

biodiversity. The number and variety of organisms found within a specified geographic region.

biological deserts. A description sometimes applied to plantations where only a single species is grown, providing impoverished habitat for a diversity of native flora and fauna.

boreal forests. Occur only in the Northern Hemisphere, mainly between 50°N to 70°N latitude. They are composed of both deciduous and evergreen species adapted to very low winter temperatures and have branches efficient at capturing light at low sun angles and shedding snow.

buffer strips. Areas of undisturbed vegetation purposely left along the borders of streams, lakes, or roads.

carbohydrates. Products of photosynthesis that contain carbon, hydrogen, and oxygen, the essential building blocks of plant structure and substrate for *plant respiration.*

carbon balance. The net change in units of carbon (molecular weight 12) for a specified system (e.g., a forest, or larger unit of land, including continents, or the whole earth) over a specified period.

catchment. The area of land where water is collected and from which it flows.

cellulose. An insoluble substance that is the main constituent of plant cell walls. It is a polysaccharide consisting of chains of glucose monomers.

clearcut. A logging practice where all trees present on an area are harvested at the same time. This is the term used to denote the practice in the United States.

clear-felling. The term used in Australia (and possibly elsewhere) to describe *clearcutting.*

climate. The average weather conditions over seasons and years.

clone. An asexually produced organism or cell from one ancestor or stock. Clones are genetically identical.

current annual increment (*CAI*). The net amount of stemwood produced by a forest in a given year. Units: $1 \text{ m}^3 \text{ ha}^{-1} \text{ yr}^{-1}$ (Conversion to imperial units: $1 \text{ m}^3 \text{ ha}^{-1} \text{ yr}^{-1} = 14.3 \text{ ft}^3 \text{ acre}^{-1} \text{ yr}^{-1}$; $1 \text{ ft}^3 \text{ acre}^{-1} \text{ yr}^{-1} = 0.067 \text{ m}^3 \text{ ha}^{-1} \text{ yr}^{-1}$).

dew point. The temperature of air or a surface at which water vapor condenses.

discounting. The economic process of determining the present value of a payment or a stream of payments to be received in the future. Future value = present value x $(1+$ interest rate$)^n$, i.e., FV = $PV(1+r)^n$, where n is the number of years. See *future value*.

drainage. The process by which water moves through layers of soil to enter streams and groundwater.

ecological stability. The capacity of ecosystems to maintain their essential functions and processes and retain their biodiversity in full measure over the long term.

economies of scale. The reduction in cost per unit from increased production, realized through operational efficiencies.

effective precipitation. The amount of precipitation that actually reaches and penetrates the soil, i.e., does not evaporate or run off.

epicormic shoots. Shoots that emerge from buds under the bark, which would probably never develop unless stimulated by fire damage to the upper parts of the tree. Epicormic shoots are characteristic of eucalypts and occur in many woody species, but not in conifers.

ethanol. Ethyl alcohol, produced from plant material and widely used to supplement petroleum fuel.

evaporation. The process by which liquid water is converted to water vapor and transferred to the atmosphere from wet surfaces (leaves, stems, litter, soil, etc.).

evaporative demand. The combined effect on the potential rate of evaporation from a canopy or the ground surface of absorbed *solar radiation*, wind, and the *humidity deficit* of the air.

evergreens. Trees that retain some foliage though all seasons.

exotic species. Foreign (i.e., nonnative or alien) to the environments in which they are planted or, in the case of insects or pathogens, introduced.

externalities. Economic costs not reflected in the assessed value of a product or service.

feedback effects. Generated by interactions between processes that may decrease (negative feedback) or increase (positive feedback) the overall effect or direction of the resultant process. For example (positive feedback): the loss of Arctic ice as a result of global warming causes the Arctic Ocean to absorb more solar radiation and so warm even faster.

field capacity (*FC*). The upper limit of the amount of water that a soil can hold against drainage under the influence of gravity.

future value. See *discounting.*

genotype. The genetic composition of an individual organism or clone thereof.

geographical information systems (*GIS*). Computer-based technology designed to displace maps with layers of data so that relations can be visualized spatially and temporarily for real or imagined situations (e.g., used to map the effects of climate change on forest productivity and forest health, and to track outbreaks of fire, disease, land clearing, or establishment of plantations).

global climate models (*GCMs*). Mathematical models that use computers to simulate the general circulation of the earth's atmosphere and the interactions between the atmosphere, land, and water surfaces.

Global Positioning Systems (*GPS*). A navigational system involving satellites and computers that computes the time difference for signals from different satellites to reach the receiver, allowing the latitude and longitude of a receiver on earth to be determined.

global warming. Increases in atmospheric and oceanic temperatures attributed to rising concentrations of carbon dioxide, methane, nitrous oxide, and chlorofluorocarbons in the atmosphere, causing an increase in the *greenhouse effect.*

Gondwana. The name given to a supercontinent that existed between 160 and 65 million years ago, which, when it split apart, formed today's southern land masses.

greenhouse effect. Incoming solar radiation passes through the earth's atmosphere, but some (greenhouse) gases—particularly carbon dioxide, water vapor, and methane—absorb a proportion of the *longwave* (heat) *radiation* emitted from the earth's surface (see *radiative forcing*). The process is essential to maintain temperatures on

earth, but increasing concentrations of greenhouse gases are causing global warming.

gross domestic product (**GDP**). The total value of goods produced and
services provided in a country during one year.

guard cells. Control the opening of leaf stomata pores in response to
light, humidity deficits of the atmosphere, and other environmental
factors.

gymnosperms. Class of trees that bear naked seeds usually held within
cones.

hardwoods. Angiosperm (flowering) trees with water-conducting systems composed of connected vessels.

humidity deficit. The difference between the amount of water vapor
that can be held by the air at a given temperature compared with
what it actually holds.

hydrologic cycle. Describes the flow of water and water vapor involved
in the hydrologic balance of a forest stand or watershed.

hydrologic equation. Expresses the change in soil water content for a
time interval (Δt) as soil water content at a given time t + precipitation – intercepted water (evaporated) – runoff – drainage out of the
soil – water transpired by plants.

isohydric. The prefix *iso* implies equality. Isohydric species maintain
their state of hydration at (approximately) some equilibrium condition by closing their stomata as soon as rates of water loss by
transpiration exceed rates of supply by movement to roots and up
through the plant to the leaves. The penalty for the isohydric response is reduction in rates of CO_2 absorption.

insolation. The amount of solar radiation received on a given surface in
a given time period. In particular, daily insolation is the solar radiation ($MJ\ m^{-2}\ day^{-1}$) received on a $1\ m^2$ horizontal surface during
one day.

Keeling Curve. The continuous record of CO_2 concentration in the
earth's atmosphere since monitoring by Charles Keeling began in
Hawaii in 1958, when atmospheric concentrations of CO_2 were at
315 ppm.

Kraft process. Chemical conversion of wood chips into cellulose to produce paper products, using sulfides.

leaf area index (LAI). The projected surface area of leaves held by vegetation per unit area of ground.

Light Detecting and Ranging Systems (LIDAR). Remote sensing technology that measures distance by illuminating a target with a laser and analyzing the reflected signal.

litterfall. The amount of leaves, twigs, and branches that fall to the ground, usually expressed as an annual rate.

long-wave radiation. Thermal radiation, characteristic of radiation from the earth, across the wavelength range 3000–5000 nm (nanometers = 10^{-9} meters), with a peak at about 10,000 nm. This is the energy that you can feel radiating from a hot body.

macroeconomics. The branch of economics concerned with the performance, structure, and behavior of a national or regional economy as a whole.

megajoules per square meter (MJ m^{-2}). A joule (J) is a unit of energy (see *watt*). A MJ is 1,000 J. MJ m^{-2} is a convenient measure of the amount of short-wave solar radiation received on a horizontal surface. MJ m^{-2} s^{-1} denotes the rate of energy receipt, or the energy intensity.

microeconomics. The branch of economics concerned with the behavior of individual consumers and firms.

monoculture. Single-species cropping system.

natural capital. The combined value to society of forests and their soil, which includes their contributions to water quality, biodiversity, carbon sequestration, and amelioration of climatic conditions.

net present value (NPV). The value of an investment today (PV) that will not mature for some years. NPV = future value/$(1 +$ interest rate$)^n$, FV = $PV(1+r)^n$, where n is the number of years.

net primary production (NPP). Biomass produced annually by plants through the process of photosynthesis, which captures CO_2 from the atmosphere, and plant respiration, which releases CO_2 back into the atmosphere.

peak outflow. A hydrological term referring to the point of the hydrograph that has the highest discharge following a precipitation event, or in response to snowmelt. Units are m^3 of water per unit of time (second, day, month, or year).

photosynthesis. The process of converting CO_2 absorbed from the air by leaves into carbohydrates, using the energy in the visible wavebands of solar radiation.

plantations. Large areas planted to single tree species for the purpose of wood production.

porosity. The ratio of the volume of pore space to unit volume of soil.

process-based models. A series of mathematical equations that express biophysical relationships in a fundamental way. Such models are sometimes called *mechanistic.*

pulpwood. Trees that are generally of smaller diameter or of lower quality than can be used to produce utility poles, lumber, plywood, or high-quality veneer.

radiative forcing. The difference between the amount of radiant energy received at the earth's surface and that radiated back to the atmosphere. Positive forcing is caused by greenhouse gases such as methane and carbon dioxide that accumulate in the earth's atmosphere and absorb long-wave radiation, reducing the amount reradiated from earth to space and causing an increase in atmospheric temperature. See *greenhouse effect.*

Reducing Emissions from Deforestation and Forest Degradation (REDD). A program aimed at reducing emissions from forest conversion and conserving carbon stocks, agreed to by some nations at the 2009 Copenhagen Climate Change Conference. This scheme relies on the world's rich nations to contribute US$10–$15 billion annually to provide economic incentives to halt deforestation by stopping illegal logging, and by promoting other means of economic development in forest-dependent communities.

relative humidity (RH). A ratio normally expressed as the percentage of water vapor held by the air compared to that at saturation.

resilience. Describes the ability of a system to absorb disturbance and still retain its basic structure and ability to function.

respiration. The process by which some of the carbohydrates in plants are broken down biochemically, releasing CO_2 back into the atmosphere. The process is universal in living organisms.

riparian zone. The area adjacent to streams or lakes that both influences the body of water and is influenced by it.

rotation. Refers, in forestry, to the interval between when a tree is planted or established by seed, and when it is harvested.

runoff. The amount of water that leaves a forest stand or watershed by running off the soil surface.

sclerophylls. Evergreen species with tough, leather-like leaves adapted to long periods of summer drought.

short-wave solar radiation (R_s). Energy from the sun, which falls within the wavelength range of 150–3,200 μm. The visible part of the spectrum—light—spans the range 400–700 μm; this is what drives the process of photosynthesis. The wavelength range from 700–2,500 μm is called the near-infrared

silviculture. The practice of controlling the establishment, growth, composition, health, and quality of forests to meet diverse needs and values.

slash. Logging debris left behind by harvesting activities.

slash and burn farming. The practice in tropical areas where all trees are cut and burned, so the ashes release nutrients to the soil, followed by cropping. Agricultural crops can usually only be grown for a few years before the soils are exhausted.

softwoods. Gymnosperm species (with naked seeds and no flowers), in which the water-conducting systems are composed of elements (tracheids) with pits on their side walls that open and close depending on the tension of the water column.

statistically significant. Widely used to indicate the probability that two sets of values are (or are not) genuinely different. The calculations are based on probability theory. The probability that observed or measured differences are real, and not the result of chance, is usually described as greater than some specified level, usually 0.95 or even 0.99; that is, the chance that the difference is *not* real is 0.05 or 0.01.

Stefan's Law. A body with a temperature greater than absolute zero (–273°C) loses heat by long-wave radiation at a rate that depends on the 4th power of its temperature.

stomata. Tiny pores in the surfaces of leaves, which open or close their apertures, allowing the uptake of CO_2 with minimum loss of water vapor.

stumpage price. The price charged by a landowner (who may be the government) to companies or operators for the right to harvest trees from that land.

sublimation. The direct transformation of snow to water vapor.

supply and demand law. In economics, *demand* means the quantity of a given article that would be bought at a given price. *Supply* means the quantity of that article that could be purchased at that price.

sustainable management. Policies and management practices that do not reduce the capacity of land to support productive growth of desired vegetation, reduce biodiversity, or create conditions likely to reduce the resources that will be available to future generations from that land.

temperate forests. Distributed mainly between latitudes 30° N to 50°N and S. They are composed of both gymnosperm and angiosperm species, including the tallest trees in the world: coast redwoods (*Sequoia sempervirens*) and mountain ash (*Eucalyptus regnans*). Temperate forests are richer in species than boreal forests but contain far fewer species than most tropical forests.

thinning. The practice of removing a proportion of the trees in a stand so that the remaining trees will grow faster and end up larger.

tipping point. Defines the place beyond which a system will move into a new state, one from which it will not likely return.

transpiration. The process by which water vapor is evaporated from the *stomata* in leaves.

tropical forests. Occur mainly within a band of about 20° of latitude around the equator where temperatures are relatively high with little seasonal or diurnal fluctuation. They represent the most biologically rich ecosystems on earth.

vapor pressure deficit (VPD). The difference between the vapor pressure of air saturated with water at a given temperature, and the actual vapor pressure of unsaturated air at the same temperature. It is expressed in units of pascals (force per unit area).

water yield. The amount of water that flows out of a catchment relative to the amount of precipitation that falls on it: e.g., if 10 mm of rain falls on a 100 ha catchment that's 10 x 100 x 10^4 = 1 x 10^7 liters (since 1 ha is $10^4 \, m^2$ and 1 mm on 1 m^2 = 1 liter). If the water yield is 60 percent, water yield is 6 x 10^6 liters. Water yield may be expressed as the ratio of outflow to input

watt. A watt is a joule (J) per second (J s^{-1}) and a joule is a unit of energy—a measure of the capacity of a system to perform work. We measure the solar energy received on a surface as watts per square meter per second (W m^{-2} s^{-1}), that is, J m^{-2}.

wilting point (WP). The moisture content of soil below which plants cannot extract water.

REFERENCES

Aber, J., N. Christensen, J. Franklin, L. Hidinger, M. Hunter, J. McMahon, D. Mladenhoff et al. 2000. "Applying Ecological Principles to Management of the U.S. National Forests." *Issues in Ecology* 6. Washington, DC: Ecological Society of America.

Battaglia, M., and P. Sands. 1997. "Modelling Site Productivity of *Eucalyptus globulus* in Response to Climatic and Site Factors." *Australian Journal of Plant Physiology* 24:831–50.

Bergh, J., S. Linder, T. Lundmark, and B. Elfving. 1999. "The Effect of Water and Nutrient Availability on the Productivity of Norway Spruce in Northern and Southern Sweden." *Forest Ecology and Management* 119:51–62.

Boyle, J. K., K. Winjum, K. Kavanagh, and E. C. Jensen, eds. 1999. *Planted Forests: Contribution to the Quest for Sustainable Species.* Dordrecht, Netherlands: Kluwer Academic.

Canadell, J. G., and M. R. Raupach. 2008. "Managing Forests for Climate Change Mitigation." *Science* 320:1456–57.

Coops, N. C., R. H. Waring, and J. B. Moncrieff. 2000. "Estimating Mean Monthly Incident Solar Radiation on Horizontal and Inclined Slopes from Mean Monthly Temperature Extremes." *Journal of Biometeorology* 44:204–11.

Dasgupta, P. 2010. "Nature's Role in Sustaining Economic Development." *Philosophical Transactions of the Royal Society, London, B Biological Sciences* 365:5–11.

Diamond, J. 2005. *Collapse: How Societies Choose to Fail or Succeed.* New York: Viking Press.

Elbakidze, M., K. Andersson, P. Angelstam, G. W. Armstrong, R. Axelsson, F. Doyon, M. Hermansson et al. 2013. "Sustained Yield Forestry in Sweden and Russia: How Does It Correspond to Sustainable Forest Management Policy?" *Ambio* 42:160–73.

Foley, J. A., N. Ramankutty, K. A. Brauman, E. S. Cassidy, J. S. Gerber, M. Johnston, N. D. Mueller et al. 2011. "Solutions for a Cultivated Planet." *Nature* 478:337–42.

Galbraith, K. 2013. "Getting Serious about a Texas-size Drought." *New York Times*, April 9.

Greenpeace. 2013. "Forest Structure." Accessed September 21, 2013. http://www.greenpeace.org/australia/en/what-we-do/forests /Forest-destruction/.

Hall, J. P. W., and D. J. Harvey. 2002. "The Phylogeography of Amazonia Revisited: Dew Evidence from Riodinid Butterflies." *Evolution* 56:1489–97.

Hamilton, L. S., N. Dudley, G. Greminge, N. Hassan, D. Lamb, S. Stolton, and S. Tognetti. 2005. "Forests and Water: A Thematic Study Prepared in the Framework of the Global Forest Resources Assessment." FAO Forestry Paper 155.

Hansen, M. C., S. V. Stehman, and P. V. Potapov. 2010. "Quantification of Global Gross Forest Cover Loss." *Proceedings of the National Academy of Sciences of the United States of America* 107:8650–55.

Henning, J. G., and D. C. Mercker. 2009. "Conducting a Simple Timber Inventory." https://utextension.tennessee.edu/publications /documents/PB1780.pdf.

Joyce, L. A., S. W. Running, D. D. Breshears, V. H. Dale, R. W. Malsheimer, R. N. Sampson, B. Sohngen, and C. W. Woodall. 2013. "Forestry." Ch. 7 in *Federal Advisory Committee Draft Climate Assessment Report Released for Public Review.* ncadac.globalchange.gov/.

Landsberg, J. J., and S. T. Gower. 1997. *Applications of Physiological Ecology to Forest Management.* San Diego, CA: Academic Press.

Landsberg, J. J., and P. Sands. 2010. *Physiological Ecology of Forest Production: Principles, Processes and Models.* Amsterdam: Academic Press.

Landsberg, J. J., and R. H. Waring. 1997. "A Generalized Model of Forest Productivity Using Simplified Concepts of Radiation-use Efficiency, Carbon Balance and Partitioning." *Forest Ecology and Management* 95:209–28.

Laurance, B. 2011. "Organised Crime, Illegal Timber and Australia's Role in Deforestation." Accessed September 21, 2013. http://theconversation.com/Organised-crime-illegal-timber-and –australias-role-in-deforestation-10048.

——. 2013. "Final Frontiers: Rainforests." Accessed September 21, 2013. https://theconversation.com/final-frontiers-rainforests-12922.

Linder, S. 1985. "Potential and Actual Production in Australian Forest Stands." In *Research for Forest Management*," edited by J. J. Landsberg and W. Parsons, 11–35. Melbourne, AU: CSIRO.

Matusick, G., G. Hardy, and K. Ruthrof. 2012. "Western Australia's Catastrophic Forest Collapse." The Conversation, May 18, 2012. http://theconversation.com/western-australias-catastrophic-forest -collapse-6925.

Millennium Ecosystem Assessment. 2005. *Ecosystems and Human Well-being: Synthesis*. Washington, DC: Island Press.

Morris, I. 2010. *Why the West Rules— for Now*. New York: Farrar, Straus & Giroux.

NASA Earth Observatory. http://earthobservatory.nasa.gov/Features /Tyndall/.

Nicholls, N. 2006. "Detecting and Attributing Australian Climate Change: A Review." *Australian Meteorological Magazine* 55:199– 211.

Nordin, A., S. Larsson, J. Moen, and S. Linder. 2011. "Science for Trade-offs between Conflicting Interests in Future Forests." *Forests* 2:631– 36.

Pan, Y., R. A. Birdsey, R. Houghton, P. E. Kauppi, W. A. Kurz, O. L. Phillips, A. Shvidenko et al. 2011. "A Large and Persistent Carbon Sink in the World's Forests." *Science* 333:988–93.

Pepke, E. "Global Wood Markets: Consumption, Production and Trade." http://www.fao.org/forestry/12711e94fe2a7dae258fbb8bc48e5 cc09b0d8.pdf.

Pittock, A. B. 2005. *Climate Change: Turning up the Heat*. London: Earthscan, CSIRO.

"Say No to Unsustainable Palm Oil." 2013. http://www.saynotopalmoil .com/.

Tans, P., and R. Keeling. 2013. Scripps Institution of Oceanography. (NOAA/ESRL) www.esrl.noaa.gov/gmd/ccgg/trends/.

Thomson, H. 2012. "The Sherwood Syndrome." http://www.aeonmagazine.com/nature-and-cosmos/hugh-thomson -britain-forests-myth.

Walker, B., and D. Salt. 2006. *Resilience Thinking: Sustaining Ecosystems and People in a Changing World*. Washington DC: Island Press.

———. 2012. *Resilience Practice: Building Capacity to Absorb Disturbance and Maintain Function*. Washington DC: Island Press.

Walter, M. E. 2010. "Earthquakes and Weatherquakes: Mathematics and Climate Change." *Notices of the American Meteorological Society* 57:1278–84.

Waring, R. H., and J. F. Franklin. 1979. "Evergreen Coniferous Forests of the Pacific Northwest." *Science* 204:1380–86.

Waring, R. H., and S. W. Running. 1998. *Forest Ecosystems: Analysis at Multiple Scales*. 2nd ed. San Diego, CA: Academic Press.

———. 2007. *Forest Ecosystems: Analysis at Multiple Scales*. 3rd ed. San Diego, CA: Academic Press.

Woods, A., J. D. Heppner, H. H. Kope, J. Burleigh, and L. Maclauchlan. 2010. "Forest Health and Climate Change: A British Columbia Perspective." *Forestry Chronicle* 86:412–22.

Young, R. 2010. "Nothofagus: Relics from Gonwana." Accessed September 18, 2013. http://www.talkingnature.com/2020/02 /biodiversity/nothofagus/.

About the Authors

 Joe Landsberg was born in Africa. He has BS and MS degrees from the University of Natal and a PhD from the University of Bristol in the United Kingdom.

His research career began in agriculture in Africa, followed by three years in Scotland working on the effects of weather on trees and ten years leading a research group in England. This led to his appointment as chief of the Division of Forest Research of Australia's Commonwealth Scientific and Industrial Research Organization (CSIRO). The Division comprised more than two hundred scientists and support staff concerned with research into all aspects of forests and forestry. Joe then moved on to become director for natural resources management in the Murray-Darling Basin Commission. In the 1990s, he worked in the Terrestrial Ecology Program of the National Aeronautic and Space Administration (NASA) and was part of the Boreas project, a major exercise aimed at improving knowledge about the effects of climate change on boreal forests. Since his retirement he has remained involved with the community of scientists and foresters concerned with forest ecology.

Joe has published three books, edited or coedited five others, and published many scientific papers. He has held visiting professorships in Australia, New Zealand, and Finland and is a member of the Finnish Academy of Science and Letters. Joe and his wife have been married for more than fifty years and have four daughters and twelve grandchildren. They live in the Blue Mountains of Australia.

RICHARD WARING is professor emeritus at Oregon State University in Corvallis. After obtaining degrees in forestry and botany at the University of Minnesota, he studied the ecology of coast redwoods while completing his PhD at the University of California, Berkeley. His first academic position, as well as his last, was at Oregon State University, College of Forestry. Dick's early research involved quantifying the environmental distribution of the flora in the species-rich forests of southwestern Oregon. From that experience, he and his graduate students created some of the first process-based predictive models of forest growth and water use.

As part of the International Biological Program in the 1970s, more complicated models were assembled but proved difficult to apply. In the early 1980s, outbreaks of bark beetles and spruce budworm provided a chance to conduct field experiments that yielded new insights about the role of insects and possible control on their populations by managing to vary tree resistance to attack. Later in the decade, NASA became interested in forests and climate change. As result of this interest, Dick and his colleagues began designing models with information acquired from earth-orbiting satellites to predict forest productivity, biodiversity, and resilience to natural agents of disturbance. In the search for common principles, scientists from other countries were drawn to visit Oregon, and in turn, Dick had opportunities to broaden his perspective through teaching and research in Europe, Australia, New Zealand, and South America.

Island Press | Board of Directors